COGNITIVE RADIO IN 4G/5G WIRELESS COMMUNICATION SYSTEMS

Edited by **Shahriar Shirvani Moghaddam**

Cognitive Radio in 4G/5G Wireless Communication Systems
http://dx.doi.org/10.5772/intechopen.74815
Edited by Shahriar Shirvani Moghaddam

Contributors

Ahmad Naeem Akhtar, Jarkko Paavola, Juha Kalliovaara, Tero Jokela, Heikki Kokkinen, Avila Jayapalan, Thenmozhi Karuppasamy, Neeta Nathani, G.C. Manna, Lutfiye Durak-Ata, Mustafa Namdar, Eylem Erdogan, Hakan Alakoca, Arif Basgumus, Shahriar Shirvani Moghaddam

First published in London, United Kingdom, 2018 by IntechOpen
IntechOpen is the global imprint of INTECHOPEN LIMITED, registered in England and Wales, registration number: 11086078, The Shard, 25th floor, 32 London Bridge Street
London, SE19SG – United Kingdom
Printed in Croatia

British Library Cataloguing-in-Publication Data
A catalogue record for this book is available from the British Library

Additional hard copies can be obtained from orders@intechopen.com

Cognitive Radio in 4G/5G Wireless Communication Systems, Edited by Shahriar Shirvani Moghaddam
p. cm.
Print ISBN 978-1-78984-706-2
Online ISBN 978-1-78984-707-9

For EU product safety concerns:
IN TECH d.o.o., Prolaz Marije Krucifikse Kozulić 3, 51000 Rijeka, Croatia,
info@intechopen.com or visit our website at intechopen.com.

Meet the editor

Shahriar Shirvani Moghaddam was born in Khorramabad, Iran, in 1969. He received his BSc and MSc degrees, both in Electrical Engineering, in 1992 and 1995, respectively. He also received his PhD degree in Electrical Engineering from Iran University of Science and Technology, Tehran, Iran, in 2001. Since 2003, he has been with the Faculty of Electrical Engineering, Shahid Rajaee Teacher Training University (SRTTU), Tehran, Iran. Currently, he is an associate professor of SRTTU and a senior member of the IEEE. In addition to more than 150 journal and conference papers, two books on digital communications, and one book chapter on MIMO systems, his research interests include cognitive radio, D2D communications, handover in cellular networks and HetNets, adaptive array processing, and DOA estimation.

Contents

Preface

The limitations of radio resources on the one hand and the rapid growth of the Internet and telecommunication applications on the other hand necessitate the optimal usage of radio resources. To effectively use available radio resources, such as time, frequency, power, and space, cognitive radio (CR) plays an integral part in future wireless communications and sensors, as well as 4G/5G non-homogeneous networks. It jointly increases throughput and balances the traffic load, either by using spectral holes in an overlay manner or by reusing the engaged resources in an underlay manner. Two decades after proposing the concept of CR by Joseph Mitola in 1998, different research works have been investigated as short-, medium-, and long-term solutions. Besides the research works focusing on spectrum sensing, spectrum sharing, resource allocation, cooperative coordinated/non-coordinated mechanisms, underlay/overly coexistence strategies, in-band/out-band scenarios, orthogonal/non-orthogonal multiple access techniques, energy-efficient algorithms, handoff/offloading protocols, and software-defined radios, a huge amount of academic and industrial interest has been shown in this field of research, especially for 4G/5G wireless communication systems and beyond. This book consists of three sections, each one including two chapters, and intends to provide the reader with an overview of the current state of the art in CR communications. The chapters cover the new research work on CR, theoretical and experimental, while introducing new ideas and suggestions for researchers for future investigations. In the first chapter, "Introductory Chapter: Primary and secondary users in cognitive radio based wireless communication systems," by S. Shirvani Moghaddam, in addition to an introduction to wireless communications and CR systems, different scenarios and examples of primary and secondary users are presented. The second chapter, "A quality of service based model for supporting mobile secondary users in cognitive radio technology," by N. Nathani et al., looks at the quality of service by focusing on availability, accessibility, maintainability, and user prediction of service. In this work, a comprehensive prediction model is employed to compute the instantaneous blocking probability on both immediate minute occupancy basis and its preceding 60 minutes basis from the time of request by secondary users. Chapter 3, "Licensed shared access evolution to provide exclusive and dynamic shared spectrum access for novel 5G use cases," by J. Kalliovaara et al., deals with the licensed shared access concept, which was initially developed to enable use of vacant spectrum resources in the 2.3–2.4 GHz band for mobile broadband through long-term static licenses. The next chapter, "Spectrum decision framework to support cognitive radio based IoT in 5G and beyond," by E.A.N. Akhtar and A.M. Siddique, aims to describe a spectrum decision support framework for CR networks and discusses how CR technology can be helpful for the IoT paradigm. The fifth chapter, "Spectrum sensing and mitigation of primary user emulation attack in cognitive radio," by A. Jayapalan and T. Karuppasamy, discusses the suitable spectrum sensing scheme for low noise environments and a trilayered solution to mitigate the primary user emulation attack, which is a severe threat to the physical layer of CR. Finally, Chapter 6, "Interference alignment in multi-input multi-output cognitive radio-based network," by A. Basgumus et al., introduces interference alignment techniques for CR networks towards 5G in order to show the demand and challenges for future wireless communications requirements.

Shahriar Shirvani Moghaddam
Associate Professor
Department of Communications
Faculty of Electrical Engineering
Shahid Rajaee Teacher Training University (SRTTU)
Tehran, Iran

Section 1

Primary and Secondary Users in Cognitive Radio Systems

Chapter 1

Introductory Chapter: Primary and Secondary Users in Cognitive Radio-Based Wireless Communication Systems

Shahriar Shirvani Moghaddam

Additional information is available at the end of the chapter

http://dx.doi.org/10.5772/intechopen.82033

1. Wireless communications and the necessity of cognitive radio

Wireless communication systems play a special role in human life, facilitate the life, and change the world much more in the future. The diversity of applications and services of wireless communications indicates the widespread use of wireless systems and equipment in the future. Since the first radio link by Marconi in 1896, it has been more than 120 years, and several wireless communication systems have been introduced up to now. Meanwhile, the advancement of cellular wireless communications, whose idea came back to 1947 and their first generation in 1978 (namely advanced mobile phone service), has been well adopted and has grown dramatically, so that the researchers and operators are focused on standardization and design of 5G and 6G wireless communication systems [1].

From the beginning, one of the main problems in wireless communications is the scarcity of radio resources (frequency, time, and power). In recent years, the lack of frequency spectrum and the necessity of frequency reuse at different times and places and reduction of the consumed power are of great importance. Looking at the following laws:

- Moore's law: The number of transistors that can be cheaply placed on a microchip doubles and the computer power exponentially increases approximately every 2 years [2]
- Cooper's law: The number of voice calls or data that can be sent over the usable radio spectrum doubles every 30 months [2]

It is clear that, in spite of the rapid technological changes, the number of requests for accessing the wireless systems and the achievable data rate is sharply increasing. In order to achieve the ultimate goal of communications, "Communications at anytime, anywhere, with anything (human, machine and object), at a low cost, low power consumption, high quality and reliability, with high speed and low latency," extensive research, design, and implementation of new techniques and technologies are required. Hence, it is necessary to anticipate the future trend in designing, upgrading, improving, and deploying wireless communication systems because it is essential to know the following:

- What are the users' requests and operators' plans?
- What are the limitations and drawbacks of current techniques and technologies?
- What is the current technology trend and what changes are needed in the future?
- What is the strategy?

In order to achieve the intended goals, the following are crucial and critical aspects [3]:

- Reusing the available radio spectrum in both time and space and use of new spectrum
- Eliminating the technological and technical constraints to achieve higher capacity
- Reducing the cost of designing, configuring, and installing a wireless network
- Decreasing the time needed to operate and deploy the system
- Reducing the hardware changes and making the ability of more software processing
- Standards, instructions, and recommendations.

One of the most important issues in wireless communications that require accurate recognition and modeling is the transmission channel. The high diversity of wireless communication channels and their time and frequency variant nature make the issue more relevant. The various types of channels in modeling, simulation, and measurements depend on the following categories and classifications:

- Indoor or outdoor/line of sight (LOS) or non-LOS (NLOS)
- Sparse or dense, based on the density of users or base stations (servers)
- The location of the users and servers to each other
- Single, double, or multilayer, in terms of coverage and the type of cells
- Homogeneity or heterogeneity of the network
- Type of control, signaling, measurement, and decision
- Noise-dominant or interference-dominant

This diversity reflects the differences and variations in the models and analyses. Therefore, it requires comprehensive research, because necessarily the model of each channel is not applicable or cannot be extended to the other channels. Hence, the dynamic cognition of the environment and the adjustment and adaptation of system parameters are the main goals.

The static spectrum allocation has over the years led to many successful applications, but it has also resulted in a situation where almost all the available frequency spectrum has been assigned to specific applications and there is no space for emerging services. On the other hand, it is shown that the spectrum is actually underutilized in time or space.

The spectrum scarcity problem has led researchers and operators to use the licensed spectrum efficiently and reuse it by considering the resource reuse techniques and introduce new frequency ranges. Briefly, we can see the main solutions for spectrum allocation:

- Array processing and signal improvement
- Multi-input multi-output (MIMO) technology
- Non-orthogonal multiple access (NOMA)
- Waveforms with high spectral efficiency, low inter-symbol interference (ISI), and a controlled peak-to-average power ratio (PAPR)
- Adaptive coding and modulation based on the signal and channel quality
- New frequency bands in the range of millimeter wavelengths and terahertz frequencies

It is suggested that new devices use the underutilized spectrum in an opportunistic manner, which is the core idea behind the cognitive radio (CR). Cognitive radio is a radio that is aware of the environment and can adapt the transmissions according to noise, interference, and channel variations. CR is based on software-defined radio (SDR). It means that a CR system at least needs "flexibility and agility", "sensing", and "learning and adaptability".

2. Cognitive radio

Cognitive radio was first introduced by Mitola [4], which is aware of the surrounding environment and is able to change its parameters to improve user performance. Measurements and research studies show that some parts of the spectrum are not used in time and space. These parts, which are named white spaces (as shown in **Figure 1**), have no active primary users. Secondary users in these parts of the spectrum will be able to detect and communicate with each other freely. This method is called the overlay manner, which is considered in many radio systems [6]. Also, in the other parts of the spectrum, the power level is very low. By using the capabilities of the cognitive radio, this part of the spectrum can be used in the underlay manner, if the transmit power of the secondary users be controlled below a predefined threshold level, and it does not make any harmful impact on the primary users [7]. Herein, we mention the coexistence of CR-based systems as follows:

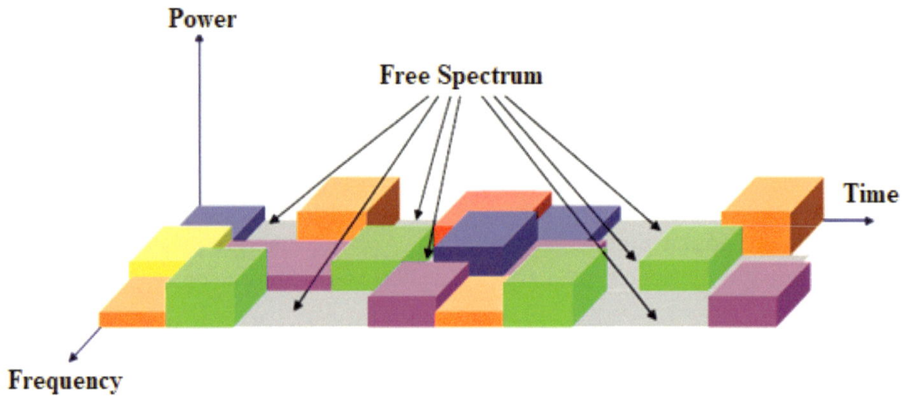

Figure 1. The free and occupied spectrums [5].

- Low-power spread spectrum communications along with existing narrowband systems
- Low-power technologies such as ultra-wide-band (UWB) and near-field communications (NFC)
- Unicast and multicast device-to-device (D2D) communications
- Multilayer heterogeneous structures based on macrocell, femtocell, nanocell, picocell, and attocell, and underlaying or overlaying conventional cellular systems.

Three important characteristics of the cognitive radio system are awareness, cognition, and adaptability [8]. Awareness means that the system has the ability to measure and sense the environment if the spectrum is free or occupied, location of the radio source, user profiles, and even the traffic and propagation characteristics of the network. Cognition, in fact, shows the ability to process information received from the environment, which should be used for making an optimal system performance. The third concept is the ability to set the parameters of the system without making any modification in the system hardware using SDR. According to these CR characteristics, it is actually a completely dynamic system in which its parameters such as frequency, transmit power, antenna pattern, transmission protocol, modulation type, coding rate, and communication technology are reconfigurable [9]. Therefore, dynamic access to the spectrum, as the main part of a cognitive radio, can help to overcome the spectrum scarcity and reuse the unused spectrum [10].

In addition to dynamic access to the spectrum, other applications are also noteworthy. For example, localization, radio frequency energy harvesting (RF-EH), D2D communications, navigation of vehicles such as a CR-based unmanned autonomous vehicle (UAV), vehicle-to-vehicle (V2V) communications, and machine-to-machine (M2M) communications.

3. Spectrum sharing in cognitive radio

Spectrum sharing, as the main part of a cognitive radio, has five important parts [11–15]:

http://dx.doi.org/10.5772/intechopen.82033

1. Spectrum sensing dependent on the following items:
    - Wireless channel: Quality of channel (path loss and noise), temporal and spatial variations of channel (multipath fading and shadowing), and type of link (LOS or NLOS)
    - Network: Homogeneous or non-homogeneous
    - User: Personal, social, and physical characteristics
    - Other aspects: Complexity, power consumption, type of management,…
2. Spectrum allocation
3. Spectrum access
4. Handshake between transmitter and receiver
5. Spectrum mobility

Spectrum sharing classifications are based on:

1. Frequency band
    - Horizontal (open) such as industrial, scientific, and medical (ISM) band
        - Coexistence
        - Spectrum sharing games
        - Centralized spectrum coordination
    - Vertical (hierarchical)
        - Reuse of TV bands
        - Spectrum pooling and common control
        - Operator-assisted
        - Spectrum load smoothing
2. Access to the spectrum
    - Underlay, such as orthogonal frequency division multiplexing (OFDM), UWB, and spread spectrum
    - Overlay (opportunistic)
3. Network structure and control scenario such as centralized, decentralized, and distributed
4. Type of cooperation of secondary users such as cooperative, non-cooperative (selfish), coexistence, spectrum sharing games, and centralized spectrum coordination

4. Primary and secondary users in cognitive radio-based systems

By defining the priority in the usage of the assigned radio frequency spectrum, two types of users, namely primary users (PUs) and secondary users (SUs), can coexist under a specific policy. Usually, the SUs are allowed to use the spectrum in such a way that they do not cause any harmful interference on the PUs, overlay or underlay. One of the important functions of a CR system in the overlay manner is spectrum sensing. It is the ability to detect the existence of the PUs in the frequency band. Many attempts have been done to develop the applicable approaches in the spectrum sensing process. Some of the centralized and distributed scenarios for narrowband and wideband signals have been introduced [16]. Complexity, signaling and overhead, presence of multiple SUs, small and large-scale fading phenomena and shadowing, and power consumption are the main aspects of designing efficient spectrum sensing algorithms. In this area, several spectrum sensing techniques such as energy detection (ED), cooperative detection, wavelet detection, and covariance detection have been proposed and investigated [17]. Spectrum sensing approaches can be categorized into blind, semi-blind, and non-blind classes. Though some methods can acquire channel state information by minimizing the variance of channel or noise uncertainty, they are not complexity-efficient for some applications such as wireless sensor networks (WSNs) [18, 19], OFDM-based cognitive radio [20], ultra-dense networks (UDNs), D2D communications, and cognitive radio heterogeneous networks (CR-HetNets). In the following, two examples of CR-based systems including primary and secondary users are presented.

4.1. CR-based HetNets

In recent years, several new technologies have been envisioned to fulfill the increasing demands of future wireless networks and meet the requirements of the fifth generation (5G) mobile standard. The main goal of these technologies is to improve the network performance in terms of the quality of service, the number of users, coverage, data rate, spectral efficiency, and end-to-end latency. A heterogeneous network (HetNet) is an important emerging technology that has been proposed for next-generation cellular networks, where multiple network tiers, from macrocells to small cells, coexist in the same coverage area [21]. Recently, CR technology has been proposed to be combined with HetNets to further improve the spectral efficiency of these networks [22]. Employing the CR technology in HetNet (namely CR-HetNet) has introduced the concept of vertical spectrum sharing in which low-priority users in a macrocell, which are referred to as cognitive secondary users (CSUs), are handed over to small cells, and they try to access the spectrum of the primary users in that cell [23, 24]. This spectrum sharing can be performed in an underlay or an overlay manner [25]. In the underlay spectrum sharing method, the CSUs can access the spectrum by only complying with the PU's stringent requirements, such as interference avoidance rules and maximum allowable transmission power. In the overlay spectrum sharing, the CSUs can utilize a portion of the frequency band left completely idle by the PUs. In a CR-HetNet, both the PUs and CSUs can be served by either the macro base station (MBS) or femto base stations (FBSs). However, as the CSU may move out of the coverage area of the MBS or the MBS may require offloading, the CSU needs to perform vertical handover (VHO) from the MBS to the FBS. In this case, the CSUs must sense the underutilized spectrum of the FBS to find an appropriate channel and then perform VHO on that channel. Once a PU requests the same channel, the CSU should leave that channel, perform spectrum handover (SHO), and switch to another appropriate free channel [26]. A management strategy is then necessary to control the number of SHOs, decrease the number of unnecessary handovers, and minimize the effect of multiple interruptions [27–30].

The CR-HetNet technology has shown great potential to meet the demands of both users and networks in the future wireless networks [31]. Hot topics in this field of research are:

- Resource allocation
- Offloading and handover due to traffic load or propagation conditions [32, 33] in the form of VHO, horizontal handover (HHO), and SHO
- Clustering to speed up the resource allocation and power control
- Power control, interference alignment, and energy harvesting

4.2. D2D communications

One of the main strategies for reusing radio resources is D2D communications between two adjacent users. D2D pairs, directly or through a relay, overlay, or underlay or in-band or out-band, transmit and receive the signal [34, 35]. By using D2D communications, the delay and the traffic load of the main network decrease, the spectral efficiency increases, and the consumed energy and cost reduce [36]. In order to do the optimum resource allocation in D2D communications, we have two approaches. In the first one, regarding the signal quality, minimizing the outage probability and in the second one, in view of the traffic and the desired throughput, maximizing the throughput can be achieved. In an in-band underlay manner, in addition to considering the noise effects and multi-path fading phenomenon, it is necessary to add the effect of interference arising from the reused radio resources [37, 38]. With the help of fixed or mobile relay nodes, the distance between the transmitter and receiver nodes of a D2D pair increases and the quality of service (QoS) improves. In the field of the D2D communications underlaying cellular network, first the communication mode should be selected between conventional cellular, relay-assisted cellular, direct D2D, and relay-assisted (or relay-aided) D2D modes. The relay has a limited coverage area much lower than a BTS, but it can handle the signals between two distant D2D users, which improves the total achievable throughput especially for the users near to the cell boundary. Several relaying strategies have been proposed, including amplify-and-forward (AF) and decode-and-forward (DF). Joint resource allocation and power control of direct/relay-aided D2D communications underlaying/overlaying cellular network is an NP-hard problem. In addition to optimum mode selection, resource allocation, and power control, finding the proper power splitting (PS) factor of relay nodes in an energy-harvested D2D communication is a goal. Achieving low-complexity closed-form expressions for the outage probability and throughput helps us to do the optimization problem in an acceptable processing time [39]. **Table 1** summarizes the advantages and disadvantages of different underlay and overlay scenarios in D2D communications, in view of interference, collision, outage probability, reuse factor, and achievable throughput.

No.	Scheme	Advantages	Disadvantages
1	Underlay (non-orthogonal)	• Activating D2D pairs with active licensed resources • No spectrum sensing needed	• Interference • No use of idle licensed resources
2	Overlay (orthogonal)	• Activating D2D pairs by idle licensed resources • No interference • Moderate spectrum efficiency	• Collision
3	Underlay/overlay	• Activating D2D pairs by both idle/active licensed resources • High spectrum efficiency	• Interference • Collision
4	Random access k-reuse underlay/overlay	• High spectrum efficiency • Higher reuse factor to 3 • Higher throughput to 3	• Interference • Collision • Higher outage to 3
5	Overlapped access k-reuse underlay/Overlay	• High spectrum efficiency • Higher reuse factor to 3 • Higher throughput to 4 • Lower outage to 4	• Interference • Collision

Table 1. Comparison of different resources reused in D2D communications [40].

Sometimes in a network, a number of nearby users may request a file (for example, a movie). In such cases, we can put those users who need the same information in a group by clustering. In the multicast D2D communications, D2D users are assigned to different clusters, each one including one cluster head and multiple cluster members. In this model, the main problem is the interference between the co-channel D2D and cellular users, which can be controlled by an efficient clustering and resource allocation. The user who can support more files will be chosen as the transmitter that multicasts the required information to the other users in that cluster. A multicast scenario combines the benefits of both a fixed cellular infrastructure and the flexibility of an ad-hoc network. In order to choose the proper cluster head and cluster the devices, it is necessary to consider some of the social and physical features. For example, social trust is a good standard for family members, friends and colleagues, and mutual trust is a good metric for beneficial collaboration among unknown users. The greater the social association of the users equals the higher chance of the nodes being in a similar group. Moreover, users close to each other need less energy for direct D2D communications, and those who have more neighbors have a higher priority for clustering. The weight of each criterion depends on the purpose of clustering [41, 42]. Hence, minimizing the outage probability and maximizing the throughput, are the main goals for resource allocation.

Hot topics in the field of unicast and multicast D2D communications are:

- Modeling based on the stochastic geometry, game theory, and graph theory
- Solving the optimization problems through fast tools and problem solvers

- Selection between direct/relay-based cellular and direct/relay-based D2D modes
- Selection between overlay or underlay scenarios
- Sharing between in-band or out-band frequencies
- Resource allocation and clustering
- Power control and energy efficiency
- Energy harvesting and energy saving

5. Future studies

New research topics that have been taken in recent years as strategies for the optimal use of radio resources, the further development of cognitive radio, and the improvement of the efficiency of wireless communication systems include:

- Massive-MIMO and ultra-dense networks
- Non-orthogonal multiple access techniques
- Power control based on beamforming or on-off techniques
- Green communications to save the resources and decrease the biological effects
- RF energy harvesting based on the time switching and NOMA techniques
- Energy-efficient algorithms for mode selection, resource allocation, and power control
- New services in 5G and 6G communications such as enhanced mobile broadband (eMBB), ultra-reliable low-latency communications (URLLC), massive machine type communications (mMTC), opportunistic Internet of Things (IoT) network, and software defined network (SDN) [43].

Author details

Shahriar Shirvani Moghaddam

Address all correspondence to: sh_shirvani@sru.ac.ir

Faculty of Electrical Engineering, Shahid Rajaee Teacher Training University (SRTTU), Tehran, Iran

References

[1] Goldsmith A. Wireless Communications. New York, NY, USA: Cambridge University Press; 2005

[2] Webb W. The Future of Wireless Communications. Boston, Norwood, Massachusetts, USA: Artech House; 2001

[3] Webb W. Wireless communications to 2020 and beyond. In: Emerging Technologies, INGENIA, Issue 41; Dec. 2009. pp. 17-20

[4] Mitola J, Maguire GQ. Cognitive radio: Making software radios more personal. IEEE Personal Communications. 1999;**6**(4):13-18

[5] Habibzadeh A. Improvement of handover in femtocell-based cellular cognitive radio network [Ph.D. Dissertation]. Iran: Shahid Rajaee Teacher Training University; 2017 (In Farsi)

[6] Haykin S. Cognitive radio: Brain empowered wireless communications. IEEE Journal on Selected Areas in Communications. Feb. 2005;**23**(2):201-220

[7] Habibzadeh A, Shirvani Moghaddam S. Noise calibrated GLRT-based spectrum sensing method for cognitive radio applications. In: 15th IEEE International Symposium on Signal Processing and Information Technology; Abu Dhabi, UAE; 7-10 Dec. 2015. pp. 174-179

[8] Peng H, Fujii T. Hybrid overlay/underlay resource allocation for cognitive radio net. in use mobility environment. In: IEEE Vehicular Technology Conference; Las Vegas, USA; Sept. 2013

[9] Xie R, Yu FR, Ji H. Dynamic resource allocation for heterogeneous services in cognitive radio networks with imperfect channel sensing. IEEE Transactions on Vehicular Technology. Feb. 2012;**61**(2):770-780

[10] Gur G, Alagoz F. Green wireless communications via cognitive dimension: An overview. IEEE Network. Mar. 2011;**25**(2):50-56

[11] Arslan H. Cognitive Radio, Software Defined Radio, and Adaptive Wireless Systems. Berlin, Heidelberg: Springer-Verlag; 2007

[12] Wyglinski AM, Kekovee M, Hou T. Cognitive Radio Communications and Networks. California, USA: Academic Press (Elsevier); 2010

[13] Berlemann L, Mangold S. Cognitive Radio and Dynamic Spectrum Access. West Sussex, UK: John Wiley and Sons; 2009

[14] Hossain E, Bhargava VK. Cognitive Wireless Communication Networks. Berlin, Heidelberg: Springer-Verlag; 2007

[15] Chen K-C, Prasad R. Cognitive Radio Networks. West Sussex, UK: John Wiley and Sons; 2009

[16] Shirvani Moghaddam S, Habibzadeh A. Cooperative spectrum sensing based on generalized likelihood ratio test for cognitive radio channels with unknown primary user's power and colored noise. International Journal of Sensors, Wireless Communications and Control. Sept. 2018;**8**(3):217-227. DOI: 10.2174/2210327908666180730092433

[17] Shirvani Moghaddam S, Kamarzarin M. A comparative study on the two popular cognitive radio spectrum sensing methods: Matched filter versus energy detector. American Journal of Mobile Systems, Applications and Services. Oct. 2015;**1**(2):132-139

[18] Shirvani Moghaddam S, Danaloo RJ. Radio identification of primary users based on cooperative compressed sensing. Tabriz JEE. 2018;**47**(4):1551-1561. In Farsi

[19] Shirvani Moghaddam S, Danaloo RJ. Cooperative compressed sensing for joint terminal localization and spectrum sensing. In: The 15th IEEE International Symposium on Signal Processing and Information Technology; Abu Dhabi, UAE; 7-10 Dec. 2015. pp. 203-208

[20] Farjami A, Tabataba Vakili V, Shirvani Moghaddam S. A simple relay selection algorithm in multi user OFDM based cooperative CR networks. In: International Symposium on Telecommunications; Tehran, Iran; 9-11 Sept. 2014. pp. 118-122

[21] Hong X, Wang J, Wang C-X, Shi J. CR in 5G: A perspective on energy-spectral efficiency trade-off. IEEE Communications Magazine. 2014;**52**(7):46-53

[22] Ghosh A, Mangalvedhe H, Ratasuk R. Heterogeneous cellular networks: From theory to practice. IEEE Communications Magazine. 2012;**50**(6):54-64

[23] Gur G, Bayhan S, Alagz F. Cognitive femtocell networks: An overlay architecture for localized DSA. IEEE Wireless Communications. 2010;**17**(4):62-70

[24] Bourdena A, Pallis E, Kormentzas G, Mastorakis G. Efficient radio resource management algorithms in opportunistic cognitive radio networks. Transactions on Emerging Telecommunications Technologies. 2014;**25**(8):785-797

[25] Mangold S, Berlemann L, Nandagopalan S. Spectrum sharing with value-orientation for cognitive radio. European Transactions on Telecommunications. 2006;**17**(3):383-394

[26] Serrano RB, Lv J, Thobaben R, Jorswieck E, Kliks A, Skoglund M. Comparison of underlay and overlay spectrum sharing strategies in MISO cognitive channels. In: The 7th International Conference on Cognitive Radio Oriented Wireless Networks and Communications; Stockholm, Sweden; 2012

[27] Wang LC, Wang CW. Spectrum handoff for cognitive radio networks: Reactive-sensing or proactive-sensing. In: The IEEE International Performance Computing and Communications Conference; Austin, Texas, USA; 2008

[28] Lee DJ, Jang MS. Optimal spectrum sensing time considering spectrum handoff due to false alarm in CR networks. IEEE Communications Letters. 2009;**13**(2):899-901

[29] Habibzadeh A, Shirvani Moghaddam S, Razavizadeh SM, Shirvanimoghaddam M. Spectrum handover mechanism for secondary users in cognitive femtocell HetNets. In: The 24th Iranian Conference on Electrical Engineering; Shiraz, Iran; 2016

[30] Habibzadeh A, Shirvani Moghaddam S, Razavizadeh SM, Shirvanimoghaddam M. Modeling and analysis of traffic-aware spectrum handover schemes in cognitive HetNets. Trans. ETT. Dec. 2017;**28**(12). DOI: 10.1002/ett.3199

[31] Shirvani Moghaddam S, Shirvanimoghaddam M, Habibzadeh A. Clustering-based handover and resource allocation schemes for cognitive radio heterogeneous networks. In: The 28th International Telecommunication Networks and Application Conference; Sydney, Australia; Nov. 2018

[32] Habibzadeh A, Shirvani Moghaddm S, Razavizadeh M, Shirvanimoghaddam M. Analysis and performance evaluation of an efficient handover algorithm for cognitive HetNets. International Journal of Communication Systems. Nov. 2017;**30**(16). DOI: 10.1002/dac.3315

[33] Habibzadeh A, Shirvani Moghaddam S, Razavizadeh M, Shirvanimoghaddam M. A novel handover decision-making algorithm for HetNets. In: The 15th IEEE International Symposium on Signal Processing and Information Technology; Abu Dhabi, UAE; 7-10 Dec. 2015. pp. 438-442

[34] Ghavami H, Shirvani Moghaddam S. Outage probability of device to device communications underlaying cellular network in Suzuki fading channel. IEEE Communications Letters. May 2017;**21**(5):1203-1206

[35] Ghavami H, Shirvani Moghaddam S. Outage probability for underlaying device to device communications. In: The International Symposium on Telecommunications; Tehran, Iran; 27-28 Sept. 2016. doi: 10.1109/ISTEL.2016.7881839

[36] Ghavami H, Shirvani Moghaddam S. Analysis of outage probability for in-band device to device communications underlaying cellular network. International Journal of Communication Systems. Feb. 2017;**30**(14):1-8

[37] Shirvani Moghaddam S. Outage analysis of non-orthogonal in-band relay-aided device-to-device communications in Rayleigh fading channels. Internet Technology Letters. March-April 2018;**1**(2). DOI: 10.1002/itl2.8

[38] Shirvani Moghaddam S, Ghavami H. Joint mode selection and resource allocation in device-to-device communications. International Journal of Sensors, Wireless Communications and Control. Sept. 2018;**8**(3):204-216. DOI: 10.2174/2210327908666180808115057

[39] Shirvani Moghaddam S. Outage analysis of energy harvested direct/relay-aided D2D communications in Nakagami channel. Journal of Communications Software and Systems. Nov. 2018;**14**(4):302-311. DOI: 10.24138/jcomss.v14i4.490

[40] Shirvani Moghaddam S. A novel resource reuse in device to device communications. In: The Internet Technologies and Society Conference, IADIS; Sydney, Australia; Dec. 2017. pp. 35-42

[41] Shirvani Moghaddam S, Ghasemi M. Efficient clustering for multicast device-to-device communications. In: 7th International Conference on Computer and Communication Engineering; Kuala Lumpur, Malaysia; Sept. 2018. pp. 228-233

[42] Shirvani Moghaddam S, Ghasemi M. A low-complex/high-throughput resource allocation for multicast D2D communications. In: 7th International Conference on Computer and Communication Engineering; Kuala Lumpur, Malaysia; Sept. 2018. pp. 234-239

[43] Bennis M, Debbah M, Poor HV. Ultrareliable and low-latency wireless communication: Tail, risk, and scale. Proceedings of the IEEE. Oct. 2018;**106**(10):1834-1853

Chapter 2

A Quality of Service Based Model for Supporting Mobile Secondary Users in Cognitive Radio Technology

Neeta Nathani and G. C. Manna

Additional information is available at the end of the chapter

http://dx.doi.org/10.5772/intechopen.80072

Abstract

Current wireless networks are characterized by a static spectrum allocation policy, where governmental agencies assign wireless spectrum to license holders on a long-term basis for large geographical regions. The operators claim that the spectrum bands for mobile operation are highly occupied. Even then, a significant amount of licensed spectrum remains underutilized. Cognitive radio senses the radio environment with a twofold objective: identify those subbands of the radio spectrum that are underutilized by the primary (i.e., legacy) users and providing the means for making those bands available for employment by secondary (i.e., unlicensed) users. For unlicensed communication, the Quality of Service parameters need to be considered. Quality of Service comprises of channel availability, accessibility, and maintainability. Assessment of vacant channels of licensed band in a geographical region is termed as availability. An analysis of the collected data lead to arrive at the conclusion that more than one-eighth part of resources of each band are nearly permanently vacant, which is enough to design in-band common control signaling methods for cognitive radio. Measurement result plot of vacant channels in cities with known population will help to assess availability of vacant channels for any city and hence, measurement complexity can be avoided. The strategy to occupy the vacant channels without disturbing the primary user operation is referred as accessibility (or selection). Accessibility of a channel is dependent on blocking probability (or Quality of Service) measured in duration of minutes instead of hours. Instantaneous blocking probability has been calculated based on current minute occupancy for all available channels as reference. A comprehensive prediction model is employed in the proposed work to compute the instantaneous blocking probability both on immediate minute occupancy basis and its preceding 60 min basis from time of request by SU. Validation through actual data establishes that channelized blocking probability estimation model has lower error value compared to estimation through prediction models of other researchers. It was also observed that hourly basis prediction model has constant blocking probability value during clock hour, whereas minutewise Grade of Service (GoS) prediction model addresses the local peak demand and hence leads to a stringent GoS estimation. On secondary user request for vacant channel, the cognitive radio network needs to evaluate the expected holding time of the particular Secondary User and to ensure channel maintainability (or allocation), and it shall predict that the allotted channel shall be able to provide interruption-free service for holding time duration. Minutewise channel occupancy traffic is bumpy in nature; hence, the present work predicts call arrival rate using Holt Winter's method. Also, at the instant of SU channel request, the channel allocation processor inputs all PU channel status minutewise, calculates actual mean residual lifetime (MRL) in minutes for each vacant channel and selects the channel with highest predicted free time. A simulation program runs on data collected from mobile switch of cellular network, which creates pseudo-live environment for channel allocation. The present work has compared the mean residual lifetime (MRL) method with the other researchers using probabilistic method of channel allocation and MRL method has been established as more accurate. The selection and allocation process with defined blocking probability model has been verified retrieving big data from data warehouse.

Keywords: cognitive radio, Quality of Service, blocking probability, mean residual lifetime

---

1. Introduction

The first decade of the twenty-first century belongs to a new wireless world indeed! The rapid growth of cellphones, Wireless Local Area Networks (WLANs), and recently the wireless Internet, in short, wireless communication is driving the whole world toward greater integrity with wireless communications. By 2020, two-thirds i.e. 66% of total IP traffic shall be occupied by Wi-Fi and mobile devices whereas wired devices will account for 34% of IP traffic in access network [1]. Licensed bands claim to be heavily congested but different research work shows that the channels in the form of time and frequency are still available. In future, wireless networks may face the problem to find suitable frequency spectrum to fulfill the demands of future services. To solve the problem of inefficient use of spectrum utilization, a new concept is evolved known as cognitive radio (CR) [2]. In 1999, Joseph Mitola III introduced the concept of CR. This new concept of CR which is called as intelligent wireless communications is capable of sensing its environment and dynamically accessing the technology. It adjusts according to the input variations of statistical data for: (a) very dependable communication wherever and whenever needed; and (b) efficiently utilizing the radio spectrum [3, 4]. This can be done by sensing the radio environment: (i) by finding spectrum bands which are unused by the PU (i.e., licensed user), and (ii) by allocating unused bands of radio spectrum to SU requesting service [5]. The underused frequency bands of PUs are called as in-band spectrum holes [6]. The spectrum holes can be used to allocate the channels to CR user. However, to ensure efficient communication for such unlicensed communication, the Quality of Service (QoS) parameters need to be considered. Quality of Service can be defined as a set of specific requirements provided by a network of users, which are necessary in order to achieve the required functionality of a service.

Cognitive radio (CR) concept is based on vacant spectrum in licensed band which sometimes referred to as combination of channels. In telecommunication, a channel refers either to a physical transmission medium such as wire or to a logical connection over a multiplexed medium such as a radio channel. Global System for Mobile Communication-900 (GSM-900)

http://dx.doi.org/10.5772/intechopen.80072

has been allocated an operational frequency from 890 to 960 MHz. GSM uses the frequency band 890–915 MHz for uplink (reverse) transmission, and for downlink (forward) transmission, it uses the frequency band 935–960 MHz. The available 25 MHz spectrum with 100 kHz guard band at two edges of the spectrum is divided into 124 Frequency Division Multiplexing (FDM) channels, each occupying 200 kHz as mentioned in **Figure 1**.

A large amount of information is transmitted between the MS and the BS, particularly, user information (voice or data) and control or signaling data. Depending on the type of information transmitted, different logical channels are used. These logical channels are mapped onto the physical channels (time slots). In the GSM system, a traffic channel will be made by a combination of a 200 kHz frequency channel and one of the eight time slots. For example, digital speech is carried by the logical channel called the traffic channel which during transmission can be allocated to a certain physical channel. There are two basic types of logical channels in GSM: traffic channels (TCHs) and control channels (CCHs). TCHs are used to carry either encoded speech or user data both in the uplink (UL) and downlink (DL) directions. The CCHs are used to communicate service between network equipment nodes.

Code Division Multiple Access (CDMA or Interim Standard-95) uses the frequency band 824–849 MHz for uplink (reverse) transmission, and for downlink (forward) transmission, it uses the frequency band 869–894 MHz. With CDMA, all users share the same 1.25 MHz wide carrier, but unique digital codes are used to differentiate subscribers. The codes are shared by both the mobile station and the base station and are called "pseudo-random code sequences". Base stations in the system distinguish themselves from each other by transmitting different portions of the code at a given time. In other words, the base stations transmit time-offset versions of the same pseudo-random code.

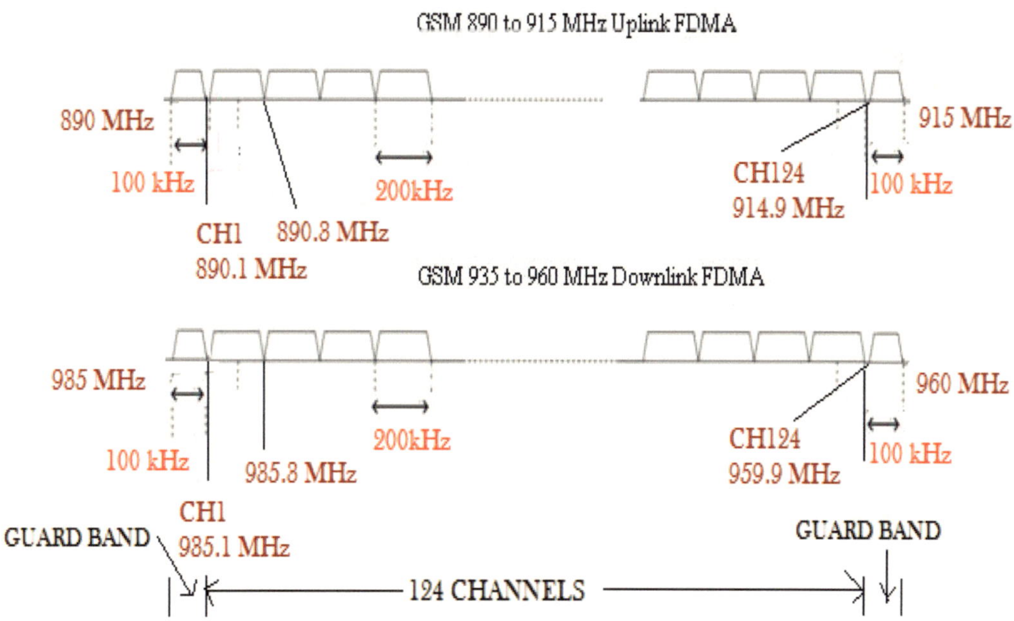

Figure 1. Frequency channels in GSM-900.

The 3rd Generation Partnership Project (3GPP) and 3rd Generation Partnership Project 2 (3GPP2) have indicated that orthogonal frequency division multiple access (OFDMA) is the choice for the physical-layer transmission technology in 4G standards. In OFDM, usable bandwidth is divided into a large number of smaller bandwidths that are mathematically orthogonal using fast Fourier transforms (FFTs). Reconstruction of the band is performed by the inverse fast Fourier transform (IFFT).

CR utilizes both licensed and unlicensed bands for communication. Among these bands, GSM bands have less attenuation; their wavelength is more resilient to phenomenon like diffraction, absorption, scattering, etc. GSM channels use FDM-TDM technique with low bandwidth of 200 kHz and hence better scalable. In practice, technologies like CDMA, OFDM, etc. uses large bandwidth and total allotted spectrum and hence, only chance to obtain large bandwidth is the un-allotted part of the licensed band. Cognitive radio users (human and machine) are low end users (users in lower or lowest economic bracket or free public utility users with minimum vocabulary or information) and expected mainly to use voice, short message and short data services. These reasons make GSM is a good choice for cognitive radio implementation.

The QoS for mobile services which has been defined by ITU-T includes different parameters of QoS like availability, accessibility, maintainability and user perception of service. These parameters have been defined in context of cognitive radio in Section 2. Availability refers to detection of unused spectrum by way of signal strength measurements. In conventional method, the signal strength of a received radio signal is measured. The measurement setup used for detection of spectrum holes in CR along with the cognitive radio issues for availability has been discussed in detail in Section 4. The proposed work calculates blocking probabilities both on immediate minute occupancy basis and its preceding 60 min basis at the instant of service request by SU. The new concept of channelized blocking probability has been defined along with the general definitions of blocking probabilities in Section 5. An algorithm has been developed to accept SU service requests with different classified Quality of Service (QoS) from a set of PU channels. Allocation of a PU vacant channel on SU call request is done based on prediction that the channel will remain vacant for more than the assessed holding time of SU. The channel allocation model works based on inputs from (a) the channel call arrival rate prediction model and (b) SU holding time assessment model and has been discussed in Section 6. The model accepts collected data as input in time serial manner for running through residual lifetime based prediction model program. The comparison of proposed work has also been done and its results and conclusion has been discussed in Section 6.

2. Quality of Service

Quality of Service (QoS) is the capability of a network to offer better service to selected network traffic over specific underlying technologies [7, 8]. The various parameters for QoS are:

i. Availability: The operator maintains a dynamic list of available channels. When the user wants to communicate, operator is liable to assign one or more communication channel to the user as per his demand and within tolerable specified time limit. In case of telecom service, this delay is maximum 6 s but usually, the delay noticed is less than a second. This function is referred to as availability.

ii. Accessibility: When the operator assigns channels to the user, the user equipment (UE) should be capable to use the allocated spectrum to the extent possible. For example, when a 200 kHz channel is allocated for some time τ, the user handset should be able to communicate at highest modulation supported by operator and RF condition. This phenomenon is called accessibility. Proper handshaking shall take place between UE and access network (AN) before establishing communication at acceptable speed by both ends.

iii. Maintainability: In mobile communication, as the user is mobile, there is a continuous change in environment and RF condition. The operator has to take into consideration various parameters like speed of communication, handover, etc. for proper maintenance of established communication. This is known as maintainability.

iv. User perception of service: It is the ability to deliver the service meeting the user's quality of expectations. It is measured by the customer satisfaction using access equipment behavior audit, drive test for mobile as pseudo customer and actual satisfaction through interrogation by customer survey specialists.

3. Availability

Over the last few years, a lot of research has undergone on spectrum sensing (SS) techniques for the detection of spectrum holes [9]. Energy detection (ED) approach, also known as radiometry or periodogram, is a popular technique for spectrum sensing due to low computational and implementation complexities [10]. The conventional SS method includes waveform-based sensing (WBS), matched filter-based sensing (MFBS) and cyclostationary-based sensing (CBS). WBS is a coherent method that correlates the received signal with the previous patterns available in database [11]. This technique is susceptible to synchronization errors which can cause false detection of primary users [10]. MFBS is the best detecting method where the received signal is interrelated with the transmitted signal [12]. The periodic characteristics of the received signals i.e., pilot sequences, carrier tones, etc. is explored by CBS technique [13]. It requires less time to achieve high processing gain due to coherent detection. In MFBS technique, it is assumed that it has the previous information of the primary's signal. It indicates that method is not suitable in some bands as some of the communication technologies are not operating with the previous information. On the other hand, CBS is unfeasible for signals that don't show cyclostationarity properties. CBS has high computational complexity [14]. Energy-based sensing (EBS) is the easiest SS method [15, 16]. This technique does not require any previous knowledge of primary user's signal but its performance is less when noise's variance is unknown or at the higher side [17]. Energy-based sensing based on sub-Nyquist sampling shall be beneficial as per as sensing duration is concerned [18]. The performance of the EBS is characterized where the PUs reflects a constant characteristic during the sensing period as well as during the sensing period where PUs can alter their ON/OFF status, thus, affecting the spectrum sensing decision [19]. A brief comparison various SS techniques is enlisted in **Table 1** as follows [10, 20]:

Spectrum Sensing Method	Characteristics
ED	1. Low complexity.
	2. No primary knowledge required.
	3. Unreliable in low SNRs.
MFBS	1. Requires prior knowledge of waveform patterns of all primary users in the various spectrum bands.
	2. Noise variance and uncertainty makes this technique unreliable as CR devices are unable to detect transmitted signal from primary users.
	3. High power consumption
WBS	1. Requires prior knowledge of synchronization.
	2. Is susceptible to synchronization errors which can cause false detection of primary users.
CBS	1. Robustness to the uncertainty in noise power.
	2. Improves the overall CR throughput
	3. High computational complexity.

Table 1. Comparison of different spectrum sensing methods.

In mobile communication, primary user occupied channels are known to network. So, a new call is eligible to occupy any of the vacant channels. In contrast, in cognitive radio network, a dynamic spectrum management is used which shall include information about the traffic pattern of the channels occupied by primary users at an instant. Basically, a CR should characterize whether the traffic pattern is static or dynamic and based on that it should use different methods for idle time prediction before selecting a channel.

Much of the spectrum below 50 GHz is available for low-powered unlicensed use. Based on environmental variations, the utilization of the licensed band is approximately 15–85% [21]. The actual utilization of mobile communication spectrum in licensed band has not yet been taken into consideration. The variation of channel utilization for various types of cities has also not been studied. These studies may be very useful to perfectly recognize the frequency channels with no active or low occupancy so that the CR technology can be successfully deployed. Few such studies has been mentioned below:

3.1. Vacant channels in Barcelona, Spain

A spectrum measurement campaign for a frequency band of 75 MHz to 3GHz was conducted through a survey in an outdoor urban environment for a continuous period of 48 h at Barcelona, Spain [22]. The six consecutive frequency bands of 500 MHz were formulated and it was found that only 22.57% of the whole frequency range was utilized.

3.2. Vacant channels in Singapore

To find the spectrum occupancy for the frequency range from 80 MHz to 5.85 GHz, another survey was conducted at Institute for Infocomm Research's building in Singapore for 24-h over 12 weekday periods [23]. It was observed that the average utilization of frequency band was only 4.54%.

3.3. Vacant channels in Limestone Maine

At the Loring Commerce Centre, a similar survey was conducted during a normal work week for 72 h for the frequency band of 100 MHz to 3 GHz [24]. In the survey, it was found that only 17% of the average spectrum is utilized during the measurement period. The ISM bands and mobile licensed bands are partly utilized and the remaining part of the spectrum band resembles noise.

3.4. Vacant channels in India

In India, the RFs are being used for different types of services like mobile communication, broadcasting, radio navigation, satellite communication, defense communication, etc. The wireless equipment are developed and manufactured based on the spectrum utilization in the country as decided by the National Frequency Allocation Plan (NFAP). The various frequency spectrums allotted to mobile communication services is shown in **Table 2** [25, 26].

CR technology has been developed to dynamically access and release channels in licensed bands. There is a scope of getting the unutilized channels in licensed spectrum with or without having a stable infrastructure for CR. Thus it is expected that at zero cost public authorities providing public utility services may be authorized to operate over unutilized spectrum even though licensed. However, such public utility service providers are very limited. Field test should be essentially conducted for the evaluation of quasi-permanently unused channels for use of in-band common control signaling purposes.

3.4.1. Measurement setup

To dynamically measure the occupancy rate of the PUs and to calculate the quantum of vacant channels available for CR use, a measurement setup called drive test equipment is used that collects data on a moving vehicle. A motor vehicle containing mobile radio network air

Technology used	Band and Channel Width	Technology Bandwidth	*Occupancy/ Remarks
CDMA	824-849MHz; 869-894MHz; Channel B/W 1.25MHz FDD; 20 channels	25+25	50% vacant
GSM 900MHz	890-915MHz; 935-960MHz; 124 channels of 200 kHz spacing	25+25	Experimentally observed results and empirical relationship (depicted in figure 2) shows that 30% of GSM channels are vacant; whereas 12% of each RF is required to manage logical 7 channels.
GSM 1800 MHz	1710-1785MHz; 1805-1880MHz; 374 channels	75+75	60% vacant channels; actual to find
WCDMA	1920-1980MHz; 2110-2170MHz	60+60	75% vacant channels
Wimax & 4G	2500–2690MHz	190	77% vacant channels
Wimax (fixed)	3400-3600MHz	200	90% vacant channels

*As per available information.

Table 2. Licensed spectrum of various wireless technologies.

interface measurement equipment is used in the drive test. The equipment measures different types of virtual and physical parameters of mobile cellular service in a given geographical region. Data relating to the network itself is collected by drive test equipment, radio frequency scanner information, services running on the network such as voice or data services and GPS information to provide location logging. The hardware and software used in the setup includes data cable and global positioning system (GPS), digital radio frequency (RF) scanner, laptop with charger and USB hub license dongle for TEMS, engineering handsets with 4 (2G/3G) SIMs of different operators mounted simultaneously and cable terminal, cell site database and link budget, clutter diagram from Google website, MapInfo software. In the setup, data collection software is installed in the laptop where mobile set is used along with GPS. Data related to signal strength, downlink and uplink frequency etc. is collected by the mobile whereas GPS collects the data of latitude and longitude of each point. All the information is stored with its geographical locations along with their respective time and date.

3.4.2. Data acquisition for availability of vacant channels

Data was collected for spectrum utilization measurements in GSM 900 MHz band in an outdoor environment of other cities viz. Bhopal, Ranchi, Patna, Dibrugarh, Shillong & Port Blair with population in the range of 1.5 million to 6.6 million as per 2011 census. The study reveals that there is 74.19% spectrum occupancy in lower band in Bhopal, while in Ranchi it is only 52.42% as

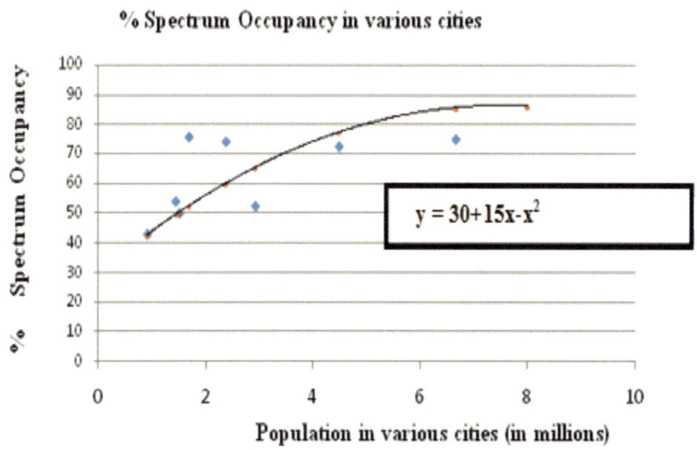

Figure 2. Population spectrum graph.

it switches to upper band where it has spectrum occupancy of 83%. In the lower band of Patna, the measurements indicate that there is 75.8% spectrum occupancy. Shillong, the capital of Meghalaya state of India is located at 25.57°N and 91.88°E on a plateau in the eastern part of the state. The population of the city is 1.43 million where spectrum occupancy is 54%. Port Blair located at 11° 40′ N and 92° 46′ E is the capital city of Andaman and Nicobar Islands in India. The next survey was conducted at Port Blair which is the municipal council in the southern part of Andaman, a part of India's Union Territory. Being the lowest populated area, it has spectrum occupancy of 43%. It is evident that most of the bands in lower band of various cities are quasi-permanently vacant. These vacant channels can be used for control signaling in CR communication.

3.4.3. *Empirical formula for channel availability*

The spectrum occupancy of eight cities of India is represented in **Figure 2**. It is shown in the diagram that there is 30% occupancy for the most sub-urban area with less population. For population between 1 million to 4 million, the increase is almost linear. In the range of population between 4 million to 7 million, it is observed that the occupancy reaches a saturation level. Also, with projected expansion of highly populated city core areas to 8 million, occupancy level is projected to reach up to 86%, leaving a clear space of 14% of channels for CR use [25].

As population increases, there will be requirement of more number of channels and this need can be managed through effective optimization methods. This can be mathematically calculated as negative requirement of channels and graphically expressed as saturation. Due to continuous growth in population, operators can request access for TCHs from higher frequency band and consequently, the occupancy at lower frequency band is reduced. Thus, the channel occupancy with growth in population can empirically be given as:

$$y = 30 + 15x - x^2 \tag{1}$$

where x = size of population in millions, y = channel occupancy percentage in lower frequency band.

Thus, it is found that 20% or more of the licensed bandwidth is almost practically unused even in a saturated market environment. In other words, more than 1/8th part of all bands were not in use which closely matches the need of one signaling channel for 7 traffic channels. This is more than the channel demand of 12% of the whole band to get access to the whole of the bandwidth at a time by cognitive radio and is adequate to take additional MAC level overhead required for CR. There is no urgent requirement of these channels by the licensed operators, whereby it can be carefully allotted as common control channel for CR purpose. Further, to doubly enhance protection of common control channel, a disaster recovery common control channel may be designated. It will hold the replica of allotments and processing status of CR Primary Common Control Channel. The above findings for common control channel are highly dynamic and sensed information may be able to provide user mobility in a most competitive environment at an economically affordable cost. The approximation of channel occupancy is possible depending upon the population and hence CR technology planners can propose a long term plan for efficient use of it for public benefits.

4. Accessibility

As per as the need of QoS is concerned, CR networks should have the ability to choose the best frequency band for use [27]. Spectrum decision is based on the channel characteristics and operations of PUs. Spectrum decision follows two steps: (i) every spectrum band is distinguished based on the statistical information of PUs and the local observations of CR users [28]. The available spectrum holes represent different characteristics that differ over time. (ii) After the available spectrum bands are characterized by considering spectrum characteristics and the QoS requirements, the most appropriate spectrum band should be selected. To minimize capacity variation, spectrum decision method is used which incorporates minimum variance-based spectrum decision (MVSD) scheme. To maximize the total network capacity, a maximum capacity-based spectrum decision (MCSD) scheme is used [29]. Accordingly a database is maintained where the data is purified based on signal to noise ratio (SNR), vacant holding time, etc. which is then used for channel allocation to SU.

After the holes are detected and best selected in licensed band, the next function of the CR user includes accessing the channel which is known as spectrum sharing. The wireless channel needs the synchronization of transmission attempts between CR users. The spectrum sharing aims to address four aspects:

a. As per the architecture, the classification can be distributed or centralized. In centralized spectrum sharing, a central entity controls the procedures of spectrum allocation and access. In distributed spectrum sharing, local or probably global policies that are independently executed by each node decide the spectrum allocation and access [30].

b. Based on allocation behavior, spectrum access can be cooperative or non-cooperative. In cooperative (or collaborative) allocation, the interference measurements of each node is exploited in such a way that it considers the effect of the communication of one node on other nodes. In non-cooperative allocation, only a single node is considered. To promote cooperation among conflicting decision makers, efficient spectrum sharing schemes such as game theory have been used for more efficient, flexible, and fair spectrum usage [31–34].

c. Based on the access technology, it is of two types: overlay and underlay. In overlay spectrum sharing, nodes access the network using the spectrum band that has not been used by PUs so as to minimize interference to the primary network. In underlay spectrum sharing, the spread spectrum techniques are exploited such that the transmission of a CR node is regarded as noise by PUs [35].

d. Spectrum sharing methods are concentrated on two types of solution: where spectrum sharing can be within a CR network, which is called intranetwork spectrum sharing; and among multiple coexisting CR networks, which is called internetwork spectrum sharing.

4.1. Daily traffic behavior

The conventional telecom uses hourly prediction for estimating mobile communication traffic. In hourly prediction, peak time is not determined neither in the beginning of the hour nor at the end of hour. The clock hour is not authentic as it does not tell exactly about the peak time at which the traffic was maximum. Hourly traffic data is insufficient to decide whether a call can be initiated at a particular instant of time. Thus, it becomes necessary to study minutewise traffic pattern. Minutewise occupancy data is computed on hourly basis for various cells of varying channel numbers, e.g., each cell with 7/14/28/60 channels for 1/2/4/8 radio frequencies (RFs) of GSM system to assess the traffic behavior of PU. There are channels with same number of RFs which are lightly loaded at some places as well as highly loaded at some other locations.

Conventionally, a telecom operator analyzes the total traffic on hourly basis and identifies the busy hour where total traffic is maximum. Hourly data arranged on weekly basis does not give a clear picture of the peak hour as it contains many peaks. The traffic variations within a clock hour are not predictable in weekly analysis. Thus, data arranged on daily basis is taken to estimate the behavior of hourly traffic. For example, the minutewise collected occupancy data was taken on hourly basis for 50 cells of different channel numbers. **Figure 3** depicts daily occupancy pattern for the locations with 60 channels (8 RFs) and differently loaded at various traffic places [36, 37]. The results are similar for other RF counts also. The figure indicates double heaps in the channel occupancy. The heap pattern shows near parabolic nature from 07:00 to 15:00 and 17:00 to 22:00 h.

Usually, peak traffic is bell-shaped around peak few minutes. Hence, the busy hour may or may not include peak traffic minute which is of serious concern for prediction of channel availability. The clock hour is not authentic as it does not exactly tell about the peak time at which the traffic was maximum. The PU busy hour is redefined in context of CR as 1 h during which peak channel occupancy occurs and calculates the growth and decay of traffic 30 min each around peak traffic minutes.

4.2. Grade of service and blocking probability

The determination of QoS provided by a particular network configuration is required for an efficient design of communication networks. The Grade of Service (GoS) is a benchmark used

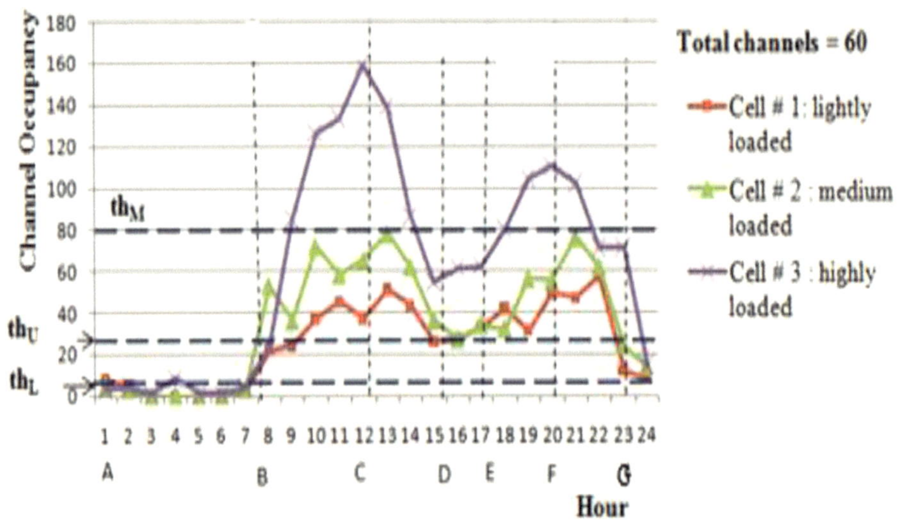

Figure 3. Daily channel occupancy pattern for three cells with 60 channels.

to define the desired performance of a particular cellular communication system by specifying a desired probability of a mobile subscriber obtaining channel access given a specific number of channels available in the system. The concept of trunking allows a large number of mobile subscribers to share the relatively small number of available channels in a cell by providing access to each mobile subscriber, on demand, from a pool of available channels.

Cellular communication systems are examples of trunked radio systems in which each mobile subscriber is allocated a channel on a per-call request basis. Upon termination of the call, the previously occupied channel is immediately returned to the pool of available channels. When a mobile subscriber requests service and in case all of the radio channels are already busy, the incoming subscriber call is blocked, or denied access to the system. In some communication systems, a queue may be used to hold the requesting mobile subscribers until a channel becomes available.

The GoS is a measure of the ability of a mobile subscriber to access a cellular system during the busiest hour. The busy hour is based upon the subscriber's demand for the service from the system at the busiest hour during a week, month or year. It is necessary to estimate the maximum required capacity in terms of available channels and to allocate the proper number of channels in order to meet the GoS. GoS is typically specified as the probability that a call is blocked, or the probability of a call experiencing a delay greater than the predefined queuing time. A call which cannot be completed at the time of call request made by a mobile subscriber is referred to as a blocked call or lost call. This may happen due to channel congestion or non-availability of a free channel. In other words, GoS is a measure of channel congestion which is specified as the probability of a call being blocked, or the probability of a call being delayed beyond a specified time.

When the offered traffic exceeds the maximum capacity of the system in terms of the allocated number of channels, the carried traffic becomes limited due to the limited number of channels.

http://dx.doi.org/10.5772/intechopen.80072

The maximum traffic is the total number of channels in Erlangs. Let us consider a cellular system that is designed for a GoS of 2% blocking. This implies that the channel allocations for cell sites are designed in such a way so that 2 out of 100 calls requested by mobile subscribers will be blocked due to channel congestion during the busiest hour.

Practically, there are two types of trunked cellular systems. The first type of trunked cellular system offers no queuing for call requests, which is known as "Erlang B" system. This means that for every mobile subscriber making a service request; it is assumed that there is no set up time to a requesting mobile subscriber. He mobile subscriber is given immediate access to a channel if it is available. If no channels are available, the requesting mobile subscriber is blocked without access to the system and is free to try again later. This type of trunking is called "blocked calls cleared or blocked calls lost". It assumes that calls arrive as determined by a Poisson distribution.

In performance evaluation of cellular systems or telephone networks, Erlang B formula is a formula for estimating the call blocking probability for a cell (or a sector, if sectoring is used) which has N "trunked" channels and the amount of ("offered") traffic is A Erlang [38]:

$$B_B\,(N, A) = \frac{\frac{A^N}{N!}}{\sum_{i=0}^{N} \frac{A^i}{i!}} \tag{2}$$

where, i = 1 to N denotes the steady-state number of busy servers. It is directly used to determine the probability B that call requests will be blocked by the system because all channels are currently used.

The second type of trunked cellular system is called "blocked calls delayed" and its measure of GoS is defined as the probability that a call is blocked after waiting a specific length of time in a queue. In this system, a queue is provided to hold the calls requested which are blocked. If a channel is not available immediately, the call request may be delayed until a channel becomes available. Customers who find all N servers busy join a queue and wait as long as necessary to receive service. The probability of a call not having immediate access to a channel is determined by the "Erlang C" formula [38]. If no channels are immediately available, the call is delayed. The Erlang C formula is expressed in terms of blocking probability as:

$$B_C\,(N, A) = \frac{\frac{A^N}{\left[N!\left(1-\frac{A}{N}\right)\right]}}{\sum_{i=0}^{N} \frac{A^i}{i!} + \frac{A^N}{N!\left(1-\frac{A}{N}\right)}} \tag{3}$$

where, N = number of trunks or service channels, A = offered load.

4.3. Limitation of Erlang traffic models

The basic unit of channel busy/idle status is recorded for each frequency and each time slot. All the channel activities (busy/idle) during each time slot and frequency correction are monitored through different counters. In addition, user friendly graphical user interface (GUI) is available from where the data can be collected and stored in backup support.

Conventionally, the data related occupancy of channel is collected on hourly basis from the counters like m15, m16, m25, m17, m18, m23, m147, and m148. Telecom occupancy related data is stored in several counters of base station controller (BSC). To get secondwise accurate data, an interrupt driven learning mechanism is required which is practically not used in telecom network because the requirement is purely academic. In earlier telecommunication, very few processors were used in the radio access logic boards. The speed of working of processors was much less as compared to present day. Further, minutewise transfer of counter data to a central computer adds to transport overhead and hence avoided. Presently, the data speed is available in processors along with high speed links. Hence, capturing of minutewise data is now feasible. Thus, the counters are read every minute for free/occupied status. A scale below minutes was not explored due to the reason that the measurement traffic is a great appreciable part of the total signaling traffic. Thus, disastrous situation cannot be introduced in a live system.

The data for individual subscriber was taken offline from Billing Center at extreme leisure hour for few subscribers. It was expected that a similar set of users shall be the SUs also. For example, the minutewise occupancy of 32 channels during a busy hour has been taken into consideration and is shown in **Figure 4** which helps to determine the availability of spectrum holes. The red color cell indicates that the channel is busy or in dedicated mode and cannot be used for channel allocation to CR. The green color cell indicates that the channel is free and can be used for CR use after ensuring its QoS parameters. The parameters like call arrival rate and user holding time of PUs is predicted for the purpose of utilization of channel by SUs.

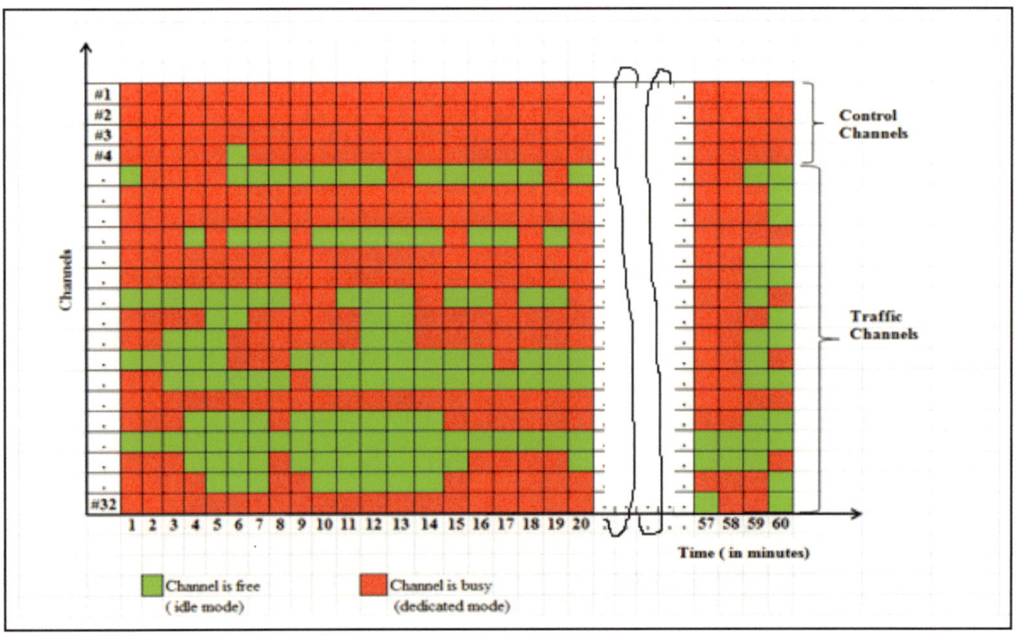

Figure 4. Minutewise occupancy chart for 32 channels in a day during busy hour.

4.4. Modified blocking probability

Blocking probability can be estimated by channel occupancy during last clock hour, e.g., 9 am–10 am at 10 am, 10 am–11 am at 11 am, etc., as in classical teletraffic theory and this estimation has been further improved through prediction models. In present chapter, clock has been considered only for hourly prediction purpose. For channel allocation, considering the instant of channel request as origin, an observation hour is defined in 2 more ways viz., (a) each hour has been composed of 60 immediately preceding minutes or channelized minutes, (b) current minute, or instantaneous minute.

For a lost call system, the GoS for CR shall be measured by using modified Poisson's model, as proposed in this chapter is given by the equation:

$$P(c, N) = 1 - \sum_{k=0}^{c-1} \frac{N^k}{k!} e^{-N} \tag{4}$$

where, $k = 0$ to $(c-1)$ with c = total number of trunked channels, $N = N_p + N_s$, N_p = count of PUs in the system, $N_s = \sigma N_p$ + offset = count of SUs in the system, where, $0 < \sigma \leq 1$. A portion of the PU, σ (known as SU factor) can be considered for the calculation of the blocking probability of a secondary call combined with PUs traffic in the system. Also, $0 <$ offset <1 such that N_s is an integer of higher value. These values of GoS help to determine whether the channel allocation to SU shall be successful or fail.

Consider a network with 'n' licensed channels ($j = 1$ to n) where the wireless nodes are static. A CRN is located within the licensed coverage area of licensed operator. The CRNs are equipped with spectrum sensor devices. The sensors monitor and report channel states to the central node via dedicated channels. Also, the outcome of the sensor state can be represented by binary signal {0,1}, where '0' represents the vacant state and '1' represents the occupied state of observed channels at an instant of time, t. All the channels are sensed assuming that the sensing time is very less than the duration of idle and busy time. The history database is periodically updated with the new sensing information. The collected database of different channels can be used to compute the different blocking probabilities as described below to estimate GoS.

4.4.1. Predicted blocking probability (PBP)

The probability computed by autoregressive moving average (ARMA) model that is a mathematical model of the persistence, or autocorrelation, in a time series is called as PBP. In ARMA model, a time series is observed for total number of calls $(y_1, y_2, \ldots. y_T)$. To predict the total number of calls in d^{th} day, forecast is done by minimizing the mean squared error (MSE), i.e., $\text{Min.}_{y'\,T+d}\, E = ((y_{T+k} - y'_{T+d})^2)$. In that case, the best forecast is the mean of y_{T+d}, conditional on the information up to T, $(y_1, y_2, \ldots. y_T)$:

$$a = y'_{T+d} = E\left(y_{T+d} \mid y_1, y_2, \ldots. y_T\right). \tag{5}$$

The BS monitoring system records the minutewise channel occupancy of licensed users for continuously 7 days of a week. The predicted value of offered load during the 8th day is calculated by using data of total calls of a particular hour for 7 days (i.e., T = 1 to 7) using ARMA model and has been depicted in **Table 3** [39]. The predicted value of total calls of 8th day of a particular hour is taken for computation of blocking probability using the formula:

$$\mathrm{PBP} = \frac{\frac{a^c}{c!}}{\sum_{i=0}^{c} \frac{a^i}{i!}} \tag{6}$$

where i = 0 to c = total channels in the system.

4.4.2. Instantaneous blocking probability (IBP)

The blocking probability provided by the system at an instant of time, (t + 1), is called as IBP. The IBP is on every minute basis as shown in **Table 4** [39]. In this case, the offered load, a, is defined as, $a = \sum_{j=1}^{n} a_j$ = number of channels busy during the minute of observation, where, $a_j = 1$, if channel is busy & $a_j = 0$ if channel is free of j = 1 to n channels at that particular instant of time. The instantaneous blocking probability at time t, is defined as:

$$\mathrm{IBP} = \frac{\frac{a^c}{c!}}{\sum_{i=0}^{c} \frac{a^i}{i!}} \text{ for } c = n \tag{7}$$

where, i = 0 to c = total channels in the system.

PREDICTION OF OFFERED TRAFFIC FOR CHANNELS j=1 to n								
HOUR → DAY ↓	1:00AM	2.00AM	.	.	.	22:00PM	23:00PM	00:00AM
#1	1243	1587	-	-	-	-	-	1325
#2	1453	1325	-	-	-	-	-	1243
-	-	-	-	-	-	-	-	
-	-	-	-	-	-	-	-	
-	-	-	-	-	-	-	-	
-	-	-	-	-	-	-	-	
#7	1000	1106	-	-	-	-	-	1106
#8*	1102*	1258*	-	-	-	-	-	1151*
*Predicted offered load								

Table 3. Prediction of offered load in a particular hour using ARMA model.

CALCULATION OF OFFERED TRAFFIC AT AN INSTANT OF TIME t+1 FOR CHANNELS j=1 to n											
TIME (in minutes) →	1	2	.	.	.	.	.	t	**t+1	.	3600
CHANNEL NO. ↓											
#1	1	0	0	1	1	0	1	1			
#2	[illegible]	[illegible]	0	0	0	0	[illegible]	[illegible]			
#3	[illegible]	[illegible]	1	1	0	1	0	0			
#4	0	1	0	0	0	1	[illegible]	[illegible]			
#5	[illegible]	[illegible]	[illegible]	0	1	0	0	0			
#6	1	0	0	0	1	0	[illegible]	[illegible]			
#7	1	1	1	0	1	0	0	1			
#8	1	1	0	0	0	0	0	0			
	**Estimation of IBP at an instant of time (t+1)										
				Active-X Command buttons for programme control							

1.INITIALIZE, 2. START, 3.Press CTRL & PAUSE, 4.IBP, 5.PROCESS, 6.CBP

Table 4. Calculation of offered load at an instant of time t = (t + 1). (a snapshot taken from software).

4.4.3. Channelized blocking probability (CBP)

The blocking probability provided by the system at an instant of time (t + 1) considering the traffic of the preceding 60 min is called as CBP and is depicted in **Table 5** [39]. The offered load in this case is defined as,

$a = \sum_{j=1}^{n} \sum_{t-60}^{t} a_t$; where, $a_j = 1$, if channel is busy & $a_j = 0$ if channel is free for j = 1 to n channels. The channelized blocking probability is defined as:

$$\mathrm{CBP} = \frac{\frac{a^c}{c!}}{\sum_{i=0}^{c} \frac{a^i}{i!}} \tag{8}$$

where, $i = 0$ to $c = 60 \times n$ = total channels in the system.

The values of CBP and IBP helps to decide the probability of success whenever a SU initiates a request. The data has been chosen at peak busy hours for 50 channels and minutewise occupancy for 300 min calls is practically taken for estimation purpose for various trunk servers ranging from 7 to 50 channels. The CBP as shown in **Table 5** can be computed by the program developed by the author.

Figure 5 is plotted for comparison of CBP and PBP for consecutive 4 h. It is evident that the standard deviation of PBP is fixed with respect to IBP but the standard deviation of CBP matches with that of IBP during the busy hour which shows that the CBP is better than PBP [39]. The CBP is much more prominent during the peak hours where random variation of instantaneous values is more.

ESTIMATION OF OFFERED TRAFFIC CONSIDERING PRECEDING 60 MINUTES DATA FOR CHANNELS j=1 to n											
Time → (in minutes)	1	2	.	.	t-60	.	.	t	**t+1	.	3600
CHANNEL NO. ↓											
#1	1	0	.	.	1	.	.	1	1		
#2	0	0	.	.	0	.	.	1	0		
.	.	.	.	.	.	.	.	.	.		
.	.	.	.	.	.	.	.	.	.		
.	.	.	.	.	.	.	.	.	.		
#50	1	0	.	.	1	.	.	1	1		
**Estimation of CBP at time (t+1) basedon immediate past 60 minutes											

Table 5. Calculation of offered load based on immediate preceding 60 min data.

4.4.4. Error estimation of blocking probability

The error is estimated by the computation of standard deviation between IBP and PBP, and IBP and CBP. The standard deviation of the sample is the degree to which individual data within the sample differ from the sample mean. Since PBP is fixed for a clock hour, the error between IBP and PBP is given by:

$$e_{pbp} = \frac{\sqrt{\sum_{i=1}^{p} (x_i - x')^2}}{p} \tag{9}$$

where x = value of IBP, x′ = predicted value of PBP with i = 1 to p = total observation, p = 60 in present case.

As CBP varies minutewise, the error between IBP and CBP is given by:

$$e_{cbp} = \frac{\sqrt{\sum_{i=1}^{p} (x_i - x_i')^2}}{p} \tag{10}$$

where, x_i' = estimated value of CBP. The present chapter proves that $e_{pbp} - e_{cbp} \geq 0$.

It is evident from **Table 6** [39] that as the number of trunk server increases, error between IBP and CBP {calculated using equation (Eq.(10))} is less than that of error between IBP and PBP

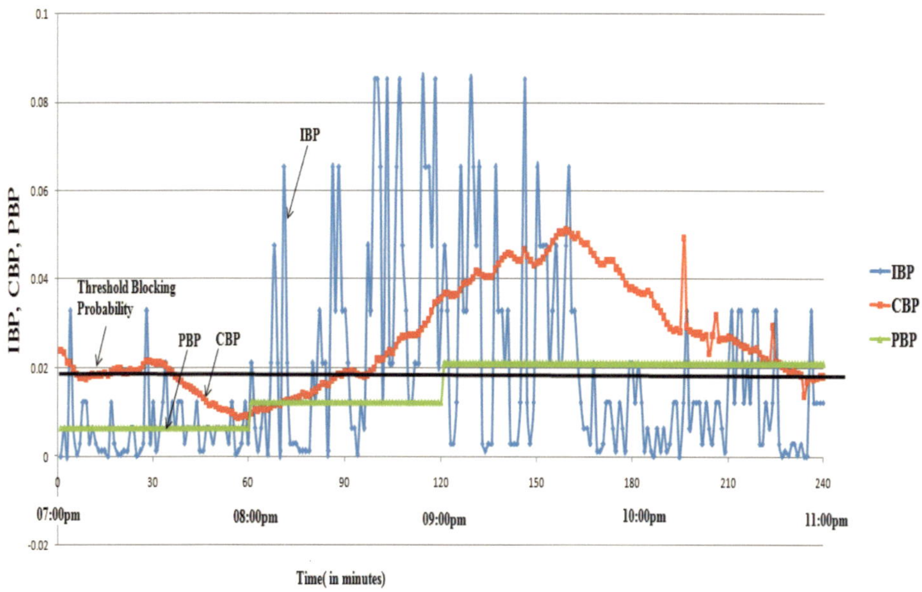

Figure 5. IBP, CBP and PBP vs. time in minutes in the system with trunk servers (c) = 22.

Trunk Servers	Std. Dev. (IBP & PBP)	Std. Dev. (IBP & CBP)	Difference
7	0.0139	0.0118	0.0021
15	0.0046	0.004	0.0005
22	0.0017	0.0018	-0.0001
29	0.0014	0.0013	0.0001
36	0.0007	0.0007	0
43	0.0003	0.0003	0
50	0.0003	0.0003	0

Table 6. Difference between standard deviation of IBP & CBP vs. IBP & PBP.

{calculated using equation (Eq.(9))}. Thus, the estimation of CBP is a better method than the estimation of PBP. The eligible list of channels available for use by SU can be formed where the channels have blocking probability ≤0.02.

4.4.5. Estimation of blocking probability at a particular instant of offer

Whenever a SU initiates a call, at an instant, the blocking probability P from Eq. (4) is measured at that instant for all channels in the cell. The value of blocking probability must be less

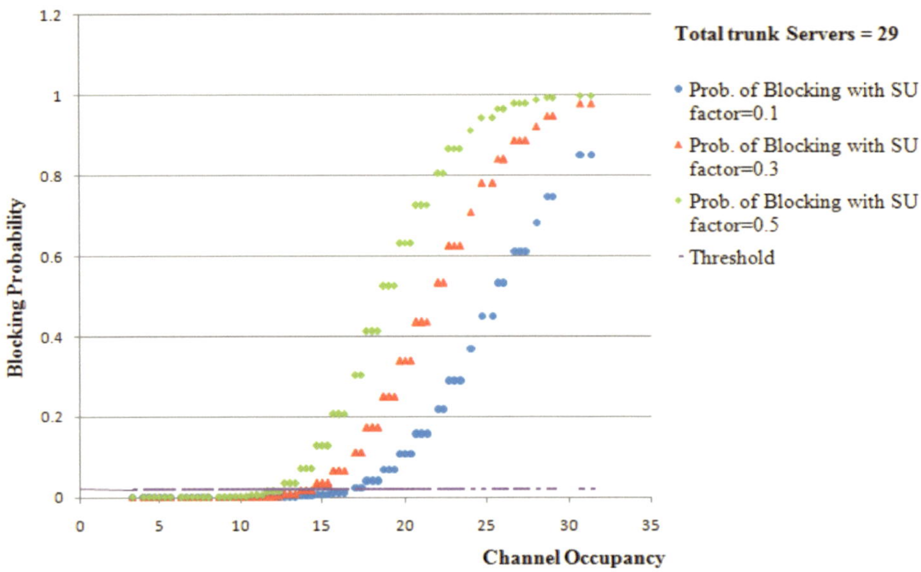

Figure 6. Call blocking probability with trunk servers (c) = 29 using Poisson's model.

than some pre-determined value. An observation for different channels was made with c = 29, 44, 60 servers to assess the blocking probability and is shown in **Figure 6** [37].

It is observed from **Figure 6** that when the primary channel occupancy <50% then the CR-BS is capable of providing mobile channel to the SU with blocking probability less than 0.02 which is equivalent to wireline. The channels which have blocking probability less than 50% are eligible used for allocation to SUs.

5. Maintainability

After capturing the best available spectrum by CR, the user may change its operating frequency band(s) that may require modifications to the operation parameters, based on the PU activity. This process is referred to as spectrum mobility. The purpose of the spectrum mobility management in CRN is to ensure smooth and fast transition that may lead to minimum performance degradation during a spectrum handoff.

5.1. Estimation of channel specific lifetime

CR users and CR infrastructure are essentially the identical as licensed authorized user system. But CR systems shall follow the guideline that: (a) only free channels of PU are to be used and when PU is active, the channels shall be returned to PU immediately, (b) it will not create any noise to PU system. Thus, architecture of SU should have some extra logic than PU system, otherwise they are similar. Therefore, there is a necessity to understand complete system architecture of PU along with PU traffic behavior for making conclusions about CR traffic handling effectively. This can be done by using prediction models.

http://dx.doi.org/10.5772/intechopen.80072

The parameters like call arrival rate and user holding time of PUs can be predicted for the purpose of utilization of channel by SUs. The probability that the channel would be accessible for a given time period is evaluated according to the prediction or estimation results. The evaluated probability is then compared with some threshold, according to which, SUs can decide whether to use this channel or not. For the purpose of prediction, the study has been arranged in two broad divisions viz. (a) daily traffic analysis for long term prediction; (b) minutewise traffic analysis for immediate prediction of availability of vacant channels.

5.1.1. Long-term prediction model

Primary channel is allotted by the network operator according to demand. The channel occupancy is recorded during each hour. The traffic pattern of each channel is seasonal in terms of daily traffic. The present study uses long term prediction model to compute call arrival rate of the PU. It takes the weekly values of call arrival rate during a particular hour as an input and predicts its weekly values for the same hour. These predicted values can be used to assess the hourly traffic of PU, based on which SU channel allocation is done.

5.1.1.1. Seasonal auto regressive integrated moving average (SARIMA) model

In SARIMA, weekly data of each cell was gathered and organized on hourly basis and one particular hour was selected and analyzed. SARIMA model was used to forecast call arrival rate of weekly data for a specific hour depending on monthly monitored data of the same hour. The prediction of traffic pattern for a week follows the following relationship:

$$\text{Forecast occupancy} = F_i = S_i{}^* \, T_j \tag{11}$$

where, i = 1 to 7 for 1 week is assumed as a seasonal unit, j = 1 to n, n = count of days for observation, S_i = seasonal coefficients; and,

$$T_i = \{A_0 - A_1{}^* D_j\}, \tag{12}$$

where, A_0 & A_1 = intercept coefficients obtained from SARIMA modeling.

5.1.1.2. Holt-Winter's (HW) method

A study of channel occupancy pattern shows that the occupancy varies every hour in a day and again daily occupancy pattern has variations over the days of a week. SARIMA uses moving average and auto regression methods which assures sample variations from predicted channel occupancy rate $\hat{\lambda}_{(j,t)}$ for jth channel at time t, as white noise. For further accurate prediction of channel occupancy rate $\hat{\lambda}_{(j,t)}$, it shall include three exponential smoothening factors viz. (a) the level (or mean) that is smoothed to give a local average value for the series of data, (b) the trend that is smoothed, and (c) each seasonal sub-series (i.e., all the values of Monday, all the values of Tuesday, etc. for weekly data) that is smoothed separately to give a seasonal estimate for each of the seasons. A combined effect of the three parameters is utilized to predict the call arrival rate by using the Holt-Winter's (HW) additive technique [40–42] given by equations as:

$$a_t = \alpha(\lambda_t - s_{t-p}) + (1-\alpha)(a_{t-1} + b_{t-1}) \tag{13}$$

$$b_t = \beta(a_t - a_{t-1}) + (1-\beta)b_{t-1} \tag{14}$$

$$s_t = \gamma(\lambda_t - a_t) + (1-\gamma)s_{t-p} \tag{15}$$

where, α, β and γ are the smoothing parameters and usually their values are chosen heuristically; a_t is the smoothed level at time t, b_t is the change in the trend at time t, s_t is the seasonal smooth at time t, p is the number of periods per season.

Here, term j is omitted from $\lambda_{(j,t)}$, and is written as λ_t for simplicity. The Holt-Winters algorithm requires starting (or initializing) values given by equations as below:

$$a_p = (1/p)\ (\lambda_1 + \lambda_2 + \ldots + \lambda_p) \tag{16}$$

$$b_p = (1/p)\ [(\lambda_{p+1} - \lambda_1)/p + (\lambda_{p+2} - \lambda_2)/p \ldots + (\lambda_{p+p} - \lambda_p)/p)] \tag{17}$$

$$s_1 = \lambda_1 - a_p, s_2 = \lambda_2 - a_p, \ldots, s_p = \lambda_p - a_p \tag{18}$$

The HW forecasts are then calculated using the latest estimates given by the equations (Eq. (13)), (Eq. (14)) and (Eq.(15) that have been applied to the series. Thus, the predicted value for $\hat{\lambda}_{(j,u+1)}$ for time period (u + 1) for jth channel as:

$$\hat{\lambda}_{(u+1)} = a_u + \tau b_u + s_u \tag{19}$$

where s_u is the smoothed estimate of the appropriate seasonal component at u, b_u is the smoothed estimate of the change in the trend value at time u and a_u is the smoothed estimate of the level at time u.

5.1.2. Short-term prediction model

The minutewise occupancy data is used as compared to hourly occupancy data that has been used in traditional traffic prediction models.

5.1.2.1. Conventional traffic prediction

The conventional telecom uses hourly prediction for estimating mobile communication traffic. In hourly prediction, peak time is not determined neither in the beginning of the hour nor at the end of hour. The clock hour is not authentic as it does not tell exactly about the peak time at which the traffic was maximum. Hourly traffic data is insufficient to decide whether a call can be initiated at a particular instant of time. Thus, it becomes necessary to study minutewise traffic pattern. In the present work, minutewise occupancy data was computed on hourly basis for various cells of varying channel numbers.

5.1.2.2. Granular traffic distribution

The nature of traffic distribution for few cells at busy hours around the peak for half an hour on both the sides with time resolution of 1 min is studied. For prediction of channel occupancy by PU, it has been established that: "The rate of change of occupancy at a particular point of time near peak time is proportional to its separation from peak time" [43]. Mathematically, it can be expressed as:

$$\frac{dy}{dt} = m' \left(t - t_p\right) \tag{20}$$

where, y = occupancy of primary channels at time t, t_p = expected time where peak occupancy occurs, m′ is an arbitrary constant.

$$\text{This evolves to}: y = at^2 + bt + c \tag{21}$$

which is the equation of a parabola with: h = –b/2a, n = ah^2 + bh + c; (h, n) are the equation of the vertex. h = $t_{p.}$

The peak of parabola may be different from peak occupancy minute. Also peak occupancy projected at peak of parabola shall be different from actual peak obtained.

5.2. Maximum duration lifetime

The authors Hao Chen and L. Trajkovic had captured 92 days (2208 h) of traffic data to study calling behavior of users [44]. They concluded that (a) time scale of minutes is too small for recording the calling activity as an average holding time of a call is usually 3–5 min, (b) time scale larger than an hour (day) is too coarse to capture. The other authors Xiukui Li and Seyed A. (Reza) Zekavat have used the concept of accessing the channel for SU which is vacant with maximum duration [45].

Hence, most of the computations in telecommunication industry are based on hourly number of calls. Accordingly, existing literatures have indicated the need for counting free lifetime of a channel as a probabilistic parameter based on hourly occupied time and hence unaware about residual lifetime, particularly in case of ON/OFF traffic channel conditions.

5.3. Mean residual lifetime (MRL)

Methods like dynamic spectrum access (DSA) are proposed to access the channel but the actual channel allocation is not taken into consideration. In case of CR, spectrum or channel mobility is the main challenge as the SU can only access a call without interfering the PU. The program in the present work shall determine the best channel eligible for allocation to SU using the concept of mean residual lifetime (MRL). The procedure for computation of MRL has been described below.

The PU channel state shall be considered as {H_o, H_1} where, H_o = free and H_1 = occupied. The primary status as sensed or predicted by the SU is shown in **Figure 7** [JAF1]. False alarm and missed detection occurs during assessment which leads either to an inefficient system or interference with PU.

Considering a small unit of time 'τ' which is the minimum time period such that BS can upload scanned RF data to Mobile Switching Centre (MSC) and fusion center without affecting routine CR operation. 'T' is the time period during which the traffic is recorded based on pulled data from different CR-BS counters and used for statistical records e.g. number of seizures of a channel per hour, total holding time of the channel per hour etc. A SU can request for a channel anytime

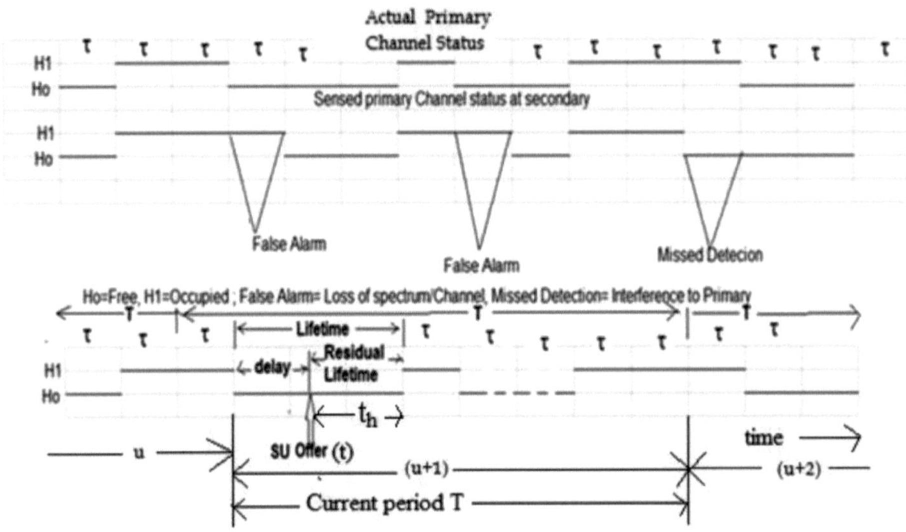

Figure 7. Channel occupancy in binary state.

within τ. The request is conveyed to the fusion center where the decision for allocation of a suitable channel is taken based on MRL and particular requesting SU's channel holding time profile.

T will be taken as an hour and (1/λ) in minutes. It is also considered that 'τ' is in minutes. It is also further considered that:

1. τ is the atomic unit of time and further decomposition of it is not practically feasible,
2. MSC is updated by BS every τ units of time,
3. MSC updates warehouse every T units of time
4. MSC updates SU traffic data in warehouse every T units of time
5. Channel occupancy request (PU&SU) is instantly passed on by the BS to MSC in real time t

These aspects will be taken up for application in different models.

Let λ is the number of calls arrived on a particular traffic data acquisition interval T. If t_h is the call holding time of a SU requesting a free PU channel at any time t, then the probability that none of the PU occupies a channel till $t = t + t_h$ is given by:

$$p_0 = \exp\left\{-\left(\hat{\lambda}/T\right) * \hat{t}_h\right\} \text{ for } t_h > \tau \,\&\, t_h > h_p, \tag{22}$$

where, $\hat{\lambda}$ = predicted call arrival rate at time t to be determined by Holt-Winter's method, h_p = predicted optimal holding (service) time, and, $\hat{t}_h$ = mean of all residual life time (t_h).

Let F be the lifetime distribution with discrete random sample and no call arrival intervals Γ_1, Γ_2, ……, Γ_n in the span of observation T. We arrange them in order such that:

$$\Gamma_1 < \Gamma_2 < \ldots\ldots < \Gamma_{kn} < \Gamma_{nn} \text{ when } \Gamma_{0n} = 0.$$

The empirical mean residual lifetime (MRL) is defined as:

$$m_n(t) = \frac{\sum_{i=k+1}^{n} (T_{in} - t)}{(n-k)} \text{ for } t_e \text{ € } \left[\Gamma_{kn}, \Gamma_{(k+1)n}\right] \tag{23}$$

and, $m_n(t) = 0$ for $t_e \geq \Gamma_{nn}$ and $k = 0,1,2,\ldots,(n\text{-}1)$.

where, $m_n(t_e)|_j$ = mean residual lifetime of jth channel which has 'n' number of vacant intervals at observation instant 't' which can be offered a SU call.

Here, t_e is the time which has elapsed since it became free; Γ_j = mean residual lifetime of jth channel at an instant t of SU call offer, and j = 1,2,…., r with r = total number of vacant channels at an instant 't' of offer.

The program developed by the author computes probability of success using the method proposed by Li and Zekavat, i.e., without MRL and with MRL. **Figure 8** depicts the probability of success with and without MRL for various trunk servers vs. time demanded by CR. The **Figure 8** clearly depicts that the proposed model using MRL method is superior than the method used by previous researchers [46].

5.4. Estimation of user holding time

A traffic model is required to represent traffic characteristics and to estimate the performance evaluation of the volume of traffic load place on network capacity and subscriber mobility. The present work assumes that the arrival rate is Poisson's distributed. The inter-arrival rate is also assumed to be exponentially distributed. The traffic model is determined assuming a certain number of channels in a cell system. The two crucial factors for mobile communication in the traffic pattern are called arrival rate of the channel and user call holding time. Hence, the traffic model is developed based on the prediction of call arrival rate and the study of holding time distribution of individual customers.

5.4.1. Poisson's distribution

A telecommunication network consists of expensive hardware (trunks, switches, etc.) which carries telecommunications traffic (phone calls, data packets, etc.). The physical network is fixed, but the traffic is random for which it is designed, i.e., the call arrival rate and the user holding time is unpredictable. Thus, to accommodate this random demand of traffic, the network designers must predict the call arrival rate for allocation of resources. The usual assumption in classical teletraffic theory is that the call arrivals follow a Poisson's process. The Poisson's assumption is consistent with data for voice traffic when the calls are generated by a large number of independently acting subscribers.

The French Mathematician Simeon Denis Poisson developed Poisson's formula. It states that for non-overlapping events, arriving at an average rate λ, the probability of 's' arrivals in time t equals:

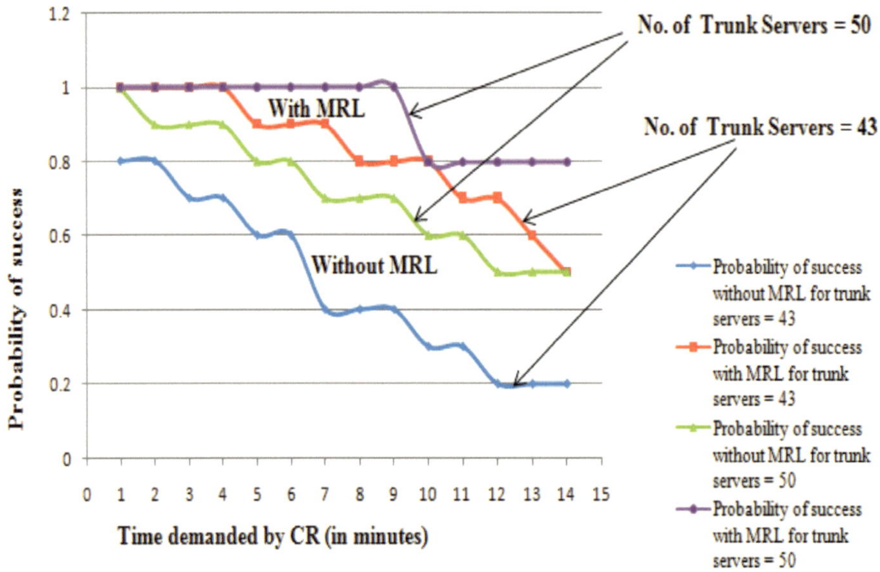

Figure 8. Comparison of probability of success with and without MRL vs. time demanded by CR for trunk servers = 43 and 50.

$$P\,(\text{'s' arrivals in time 't'}) = \frac{(\lambda t)^{s} e^{-\lambda t}}{s!} \tag{24}$$

Poisson's distribution is taken for traffic measurement as it is based on memoryless system and it generally gives a better estimate of the traffic related parameters.

5.4.2. Holding time distribution

Telecom occupancy related data is stored in several counters of Base Station Controller (BSC). To get secondwise accurate data, an interrupt driven learning mechanism is required which is practically not used in telecom network because the requirement is purely academic. In earlier telecommunication, very few processors were used in the radio access logic boards. The speed of working of processors was much less as compared to present day. Further, minutewise transfer of counter data to a central computer adds to transport overhead and hence avoided. Presently, the data speed is available in processors along with high speed links. Hence, capturing of minutewise data is now feasible.

The channel occupancy duration depends on individual person depending upon profession, status, time of day, etc. and varies widely at different hours of a day. This is most predominant in case of speech communication when the channel holding time depends upon caller and various customers called parties. Thus, the holding time data shows different variation at different levels of the time series and hence, a transformation of the data series can be useful. The Box- Cox method is used to transform this data into normality. The Box-Cox method obtains a normal distribution of the transformed data (after transformation) and a constant variance.

Let us denote original observations $h_{i,u,p}$ as h_p for the ith user at u^{th} hour and write the series as $h_1,h_2,—,h_t$ and transform the observations as $w_1,w_2,—,w_t$. According to Box-Cox principle [47]:

http://dx.doi.org/10.5772/intechopen.80072

$$\begin{aligned} w_t &= \log(h_t); && \text{for } \zeta = 0 \\ \text{and } w_t &= \left(h_t^{\zeta} - 1\right)/\zeta; && \text{for } \zeta \neq 0 \end{aligned} \tag{25}$$

where, ζ is a parameter used to compute the confidence level (CL). Using the values of $\zeta = \{-2, -1, -0.5, 0, 0.5, 1, 2\}$ and to gain confidence level (CL) limit up to 95%, the optimum value of ζ shall be used to assess h_p as:

$$\begin{aligned} h_p &= \exp(w_t); && \text{for } \zeta = 0 \\ \text{and } h_p &= (\zeta.w_t + 1)^{1/\zeta}; && \text{for } \zeta \neq 0 \end{aligned} \tag{26}$$

where, h_p = optimal holding (service) time of the user at time t.

The Box-Cox method can be used to decide the optimum holding time of the user.

5.5. Channel allocation model

On placement of service request by an SU, all vacant channels from eligible list has to be evaluated for allocation based on (i) predicted call arrival rate during the hour created through long term table; (ii) IBP at the instant for primary decision on allocation; (iii) mean residual lifetime of the channel at the time of service request; (iv) expected service holding time of a particular user requesting service; (v) CBP at the instant based on short term table. Finally, the best channel with highest probability of survival is selected for offer to the incoming SU traffic [46].

The interworking of different blocks for channel allocation is shown in **Figure 9** and is described below:

A. Channel traffic updation: The Gateways (GWs) monitor PU activities using dedicated RF scanners. $G_1, G_2, \ldots, G_k$ are responsible for monitoring PU activities as well as for SUs. Any change in channel occupancy is passed on by GW to CR-BS and MSC in real time t. BS maintains several counters for traffic recording purposes. The counts of the counters are polled by MSC every 'τ' interval and then the counters are reset. When $T = k\tau$ where, $k = 2, 3, \ldots$ MSC prepares a table for the call arrival rate for T interval for each channel and deposit to the warehouse where the data is stored in format $\lambda(j, u)$ where, j = channel number and u = current T period number. This module also provides idle time information $d_1, d_2, \ldots$, etc. in τ units since a channel is free.

B. SU traffic updation: SU traffic information is recorded in the billing register after the completion of each call. The traffic details in respect of each SU are stored in warehouse. It is used for predicting holding time of SU at the time of service request.

C. Channel traffic prediction: Primary channel is allotted by network operator according to demand. The channel occupancy is recorded during each T. The predicted value for channel occupancy $\hat{\lambda}_{(j,u+1)}$ for time period (u + 1) for jth channel is computed using HW method using (Eq.(19)).

As soon as a new predicted value of call arrival rate is available during a particular hour, HW updates its estimated three components (level, trend, seasonal) for that particular

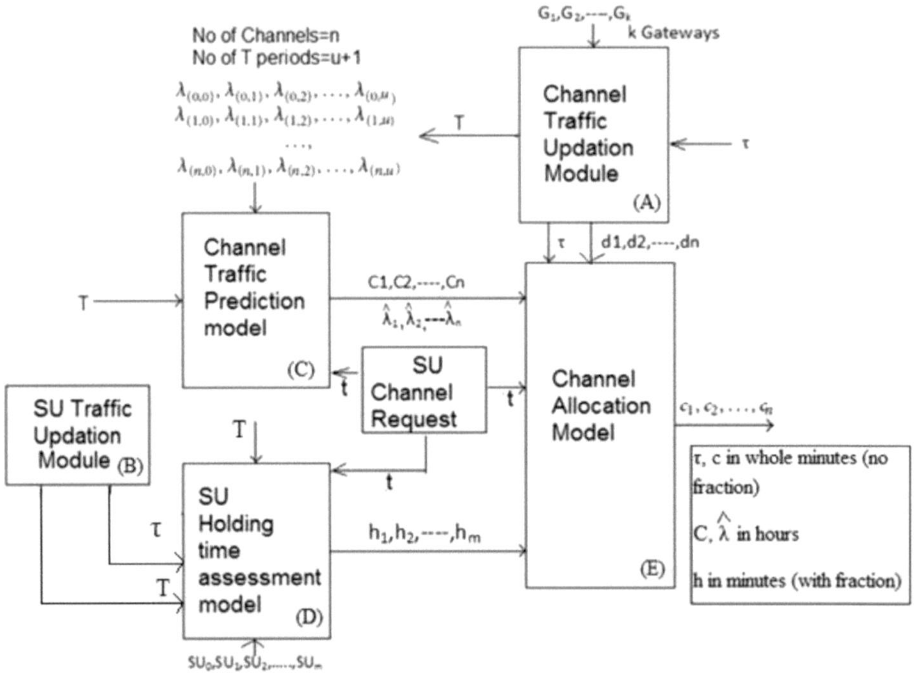

Figure 9. Channel selection block diagram.

hour. The value of smoothing constant for each component falls between zero and one. Larger smoothing constants mean more weight is placed on the value suggested by the new predicted value and less on the previous estimate. This means that the method will adapt more quickly to genuine changes in the call arrival pattern.

The list of channels $\{C_1, C_2, \ldots\}$ which satisfies the above condition are predicted during the last hour and are then transferred to the channel allocation model for assessment of holding time during the current hour of allocation.

D. SU holding time assessment: The holding time data shows different variation at different levels of the time series and hence their transformation is done by Box-Cox method using (Eq.(25)) and the optimum holding time needed by SU is obtained by (Eq.(26)).

E. Channel allocation model: The set of eligible channels obtained by (Eq.(22)) are with residual lifetime:

$$t_j = \Gamma_j - h_p > 0 \tag{27}$$

where, h_p = service time needed by SU at time t.

The eligible channel set $\{c_1, c_2, \ldots, c_r\}$ is arranged and the probability of the success in the offered jth channel shall be:

$$p_{0,j} = \exp\left[-\left\{\hat{\lambda}_{(u+1)}/T\right\} * t_j\right] \tag{28}$$

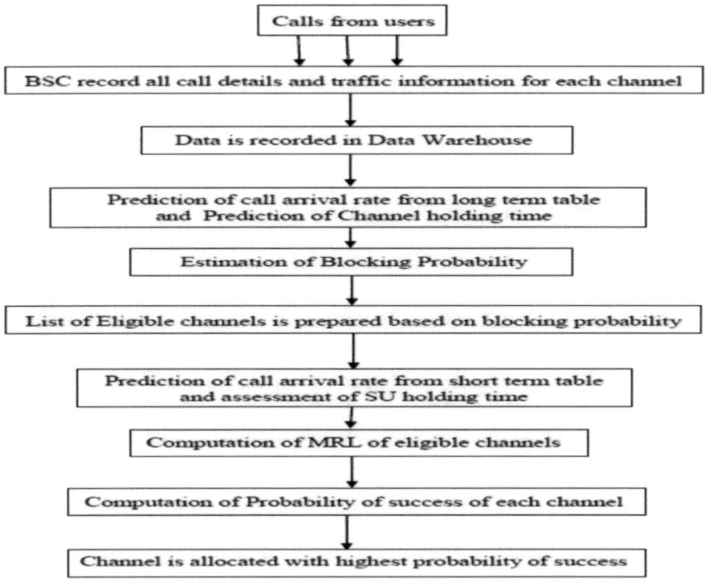

Figure 10. Process for channel allocation.

A sequential flowchart of data flow from all the blocks and computation of MRL has been given in **Figure 10**. Based on the measurement of the call arrival rate, holding time of PUs and various parameters of QoS, the channel allocation is done.

5.6. Simulation and verification of results

Minutewise traffic data acquisitioned online is collected in a table in OMC-R. SU can raise service request at any time. To serve channel allocation engine inputs,

a. At the beginning of each hour, for all channels predicted call arrival rate $\hat{\lambda}_{(j,t)}$ for the current hour from channel prediction model;

b. estimated holding time in minutes for the service request from SU holding time assessment model;

c. all channel occupancy status is 'free' or 'busy' mode for last 60 min starting from current minute from channel traffic updation module is taken, and

d. their MRL is computed.

Finally, channels with highest probability of survival given by Eq. (28) are selected for offer to the incoming SU traffic.

A trial run for 25 times each for available channels 7, 15, 22, 29, 36, 43 and 50 were carried for holding times from 1 to 14 min in the program developed by the author. A snapshot of the program is depicted in **Figure 11**, where different SUs demand varying holding times. The 'red' color indicates that the channel is busy while the 'green' color indicates that the channel is free. Whenever an SU requests a channel at time t for a certain holding time, the channel is allocated to

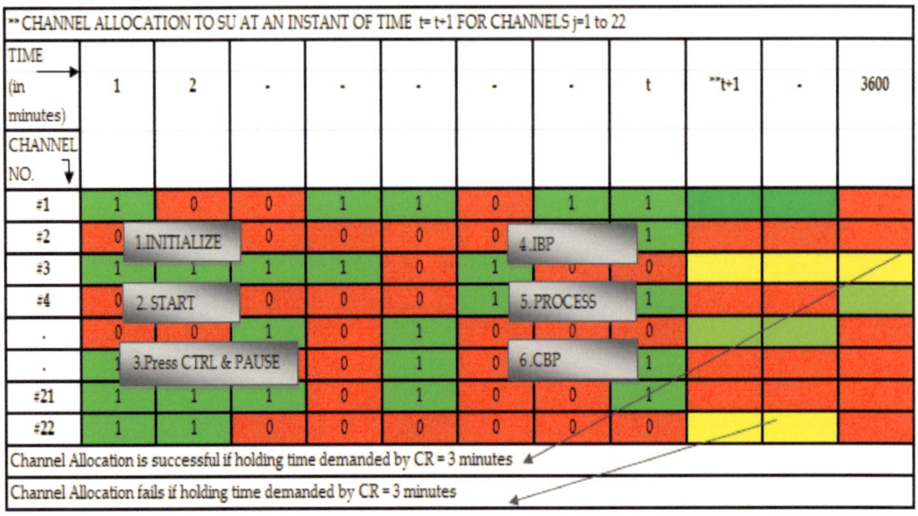

Figure 11. A snapshot of the program for channel allocation at an instant of time t.

that particular SU if the channel is free for continuous holding time demanded by SU, and is shown in 'yellow' color in **Figure 11**. If the channel is busy even for the last holding time demanded by SU, the program indicates that the channel allocation is unsuccessful and is shown with 'red' color in continuation with yellow color.

To study success rate for channel allocation various cases were considered with varying available number of channels and different holding times demanded by SU, where the total number of channels in the cell is denoted by t_{ch}.

Case (i): Consider $t_{ch} = 15$ and the time demanded by the SU = 2 min. The result is obtained when the program is initialized at 100th min and stopped at 102nd min out of 300 min total available data. At 102nd min, when a call request is made by SU, the total available vacant channels are 5 out of 15 channels. The values of blocking probabilities are obtained as IBP = 0.03649 and CBP = 0.0308. The MRL is computed for all the five vacant channels. The MRL of the five channels are 4.7, 0.82, 0.5, −27 and −27.25. As one of the channels has maximum MRL and has value > holding time demanded (2 min), it is selected for channel allocation to SU. The result is depicted in snapshot given in **Figure 12** and is shown with continuity of yellow color for 2 min, which signifies that the channel allocation was successful. **Table 7** shows analysis of success rate for 15 channels with various holding times demanded by CR.

Case (ii): Consider $t_{ch} = 43$ and the time demanded by the SU = 9 min. The result is obtained when the program is initialized at 80th min and stopped at 82nd min out of 300 min total available data. At 82nd min, when a call request is made by SU, the total vacant channels are 14 out of 43 channels. The values of blocking probabilities are obtained as IBP = 1.102×10^{-3} and CBP = 6.442×10^{-4}. The MRL is computed for all the 14 vacant channels. The MRL of the 14 channels are: 4, 0.75, 0.444444, −0.6875, −1.5, −2.6, −4, −5.66667, −5.8, −11.6667, −12.5, −13.5714, −13.8, −14.2222, −20. As the MRL of one of the channels has value > holding time demanded (9 min); thus, the channel allocation was successful. The result is depicted in snapshot given in **Figure 13** and is shown with yellow color which signifies that the channel allocation was successful. **Table 8** shows analysis of success rate for 43 channels with various holding times demanded by CR.

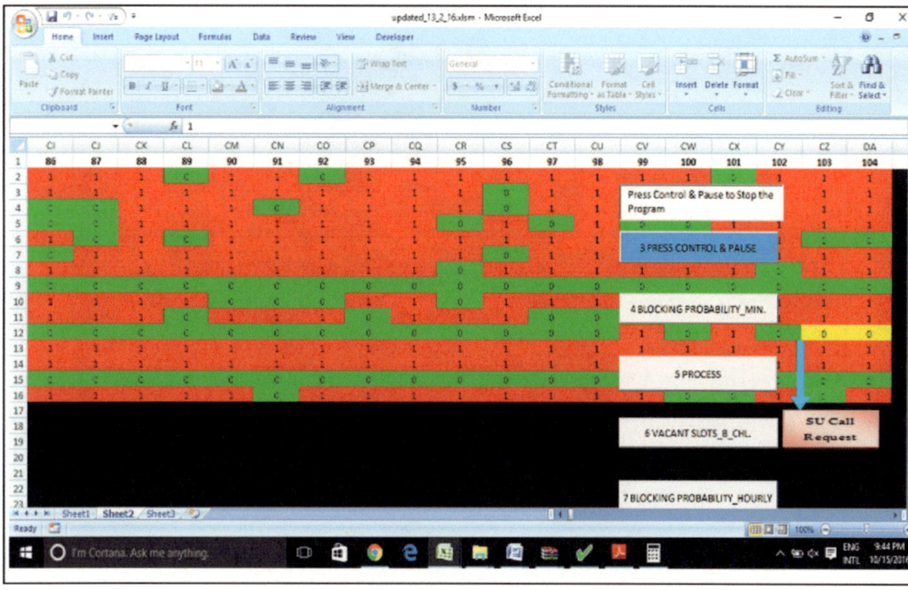

Figure 12. Validation result for total channels = 15; SU holding time demanded = 2 min.

OBSERVATION	INSTANT OF OFFER	TIME DEMANDED BY CR													
		1	2	3	4	5	6	7	8	9	10	11	12	13	14
#1	65														
#2	70														
#3	75														
#4	80														
#5	85														
#6	90														
#7	95														
#8	100														
#9	105														
#10	110														
#11	115														
#12	120														
#13	125														
#14	130														
#15	135														
#16	140														
#17	145														
#18	150														
#19	155														
#20	160														
#21	165														
#22	170														
#23	175														
#24	180														
#25	185														
SUM (SUCCESS)		20	15	7	5	6	4	1	1	1	1	1	1	1	1
PROBABILITY OF SUCCESS		0.8	0.6	0.28	0.2	0.24	0.16	0.04	0.04	0.04	0.04	0.04	0.04	0.04	0.04

Table 7. Computation of success rate for total channels = 15 for CR holding time = 1 to 14 at various runs of the program.

Thus, the channel allocation is successful as the count of channels increases in the system with the precondition that the MRL > time demanded by the SU.

5.7. Validation of result

The data was chosen at busy hours for various channels ranging from 15 to 50 and minutewise occupancy for 300 min calls is taken for simulation purpose. A program has been developed for

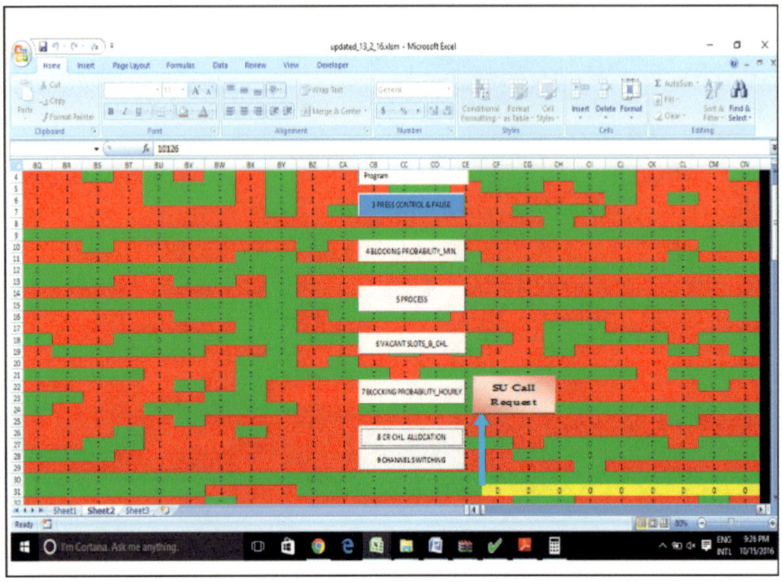

Figure 13. Validation result for total channels = 43; SU holding time demanded = 9 min.

OBSERVATION	INSTANT OF OFFER	TIME DEMANDED BY CR													
		1	2	3	4	5	6	7	8	9	10	11	12	13	14
#1	65														
#2	70														
#3	75														
#4	80														
#5	85														
#6	90														
#7	95														
#8	100														
#9	105														
#10	110														
#11	115														
#12	120														
#13	125														
#14	130														
#15	135														
#16	140														
#17	145														
#18	150														
#19	155														
#20	160														
#21	165														
#22	170														
#23	175														
#24	180														
#25	185														
SUM (SUCCESS)		25	24	21	21	20	18	18	16	16	14	13	13	10	10
PROBABILITY OF SUCCESS		1	0.96	0.84	0.84	0.8	0.72	0.72	0.64	0.64	0.56	0.52	0.52	0.4	0.4

Table 8. Computation of success rate for total channels = 43 for CR holding time = 1 to 14 at various runs of the program.

real time offering and verification if the call request succeeds or fails for the duration demanded by SU. The program calculates the predicted λ upto last hour at background and has been included in simulation program as offline. Similarly, holding time needed for SU has also been imprinted interactively. The program accepts any number of PU channels upto 50 selectively at the time of trial. The position of occupancy can be seen on screen starting from any instant after 60 min upto 300 min for any duration. Also, one or more SU calls can be offered to the system and minute by minute observation of SU call progress can be monitored on screen.

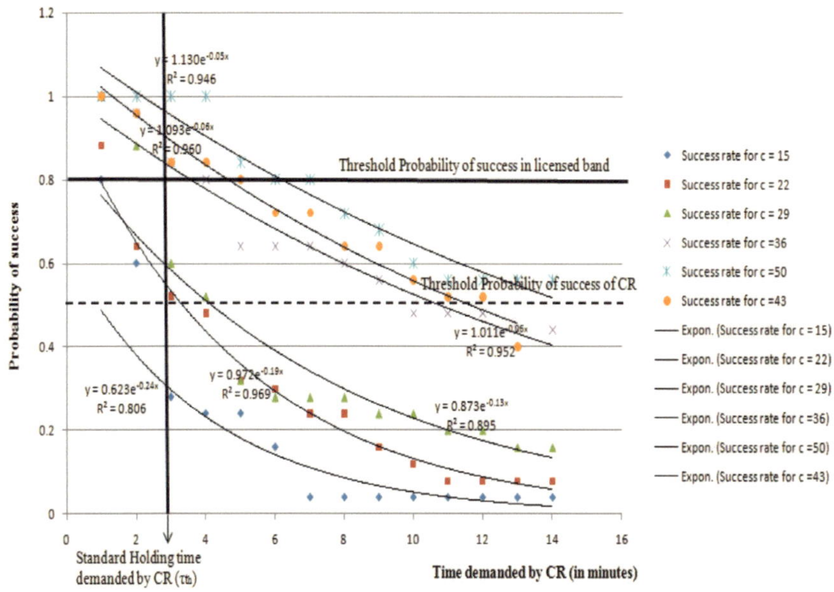

Figure 14. Probability of success rate vs. time demanded by CR for various trunk servers.

The program was run repetitively and at random, under various channel availability conditions and differently demanded holding time. The result has been plotted in **Figure 14**, where the probability of success (or QoS) has been calculated as:

$$\text{probability of success (or QoS)} = \text{number of times call success/total number of trials,}$$

where, number of trials were 25 times for each run condition at random input time.

As depicted in **Figure 14**, in case, if the threshold probability in licensed band is taken to be 0.8, the success rate for CR users is still achieved if the channels are ≥36 and holding time ≤2.5 min demanded by the SU [46]. When the threshold probability for SU is 0.5, the success rate is achieved if the channels are ≥22 with minimum holding time ≤ 2.5 min demanded by the SU. Thus, when the PU channel occupancy is 50%, the CR-BS shall provide mobile channel to SU with blocking probability ≤0.02, where industry standard for blocking is 2% (0.02).

Thus, the CR-BS shall be capable of providing vacant channels to SU. Grade of Service (GoS) of SU is at par with GoS standard specified for PU when traffic intensity is below 50%. Success rate of channel allocation is increased as number of channels increases in the system.

6. Conclusion

In mobile communication network, despite heavy usage of communication channels, vacant channels are available which can be used for cognitive radio network. From the present work it is found that practically 20% or more of the total licensed bandwidth is permanently unused or vacant even in a crowded region This is well above the need of 1 out of 8th part of the band to get access to the whole of the bandwidth at a time by CR and adequate to take additional MAC level overhead required for CR. Remaining 80% of the licensed bands are dynamically vacant which can be used for traffic purpose. Based on results, an empirical relationship has been established for channel occupancy for a city, where such survey has not been conducted but population is known. The blocking probability of the channels are available for allocation and duration of remaining unoccupied can be mapped to the instant status of the channels. The present work has established that when the PU channel occupancy is 50%, the CR-BS shall provide voice channel to SU with blocking probability ≤0.02, where industry standard for blocking is 0.02. The GoS improves linearly with total number of channels in the system at a given per channel availability. It is also evident that Erlang theory is effective for Poisson's distribution theorem with ≥20 channels for GoS to achieve.

A program that has been developed to accept SU service requests with different QoS from a set of PU channels can be used for dynamic allocation to SU. A SU call request can be placed in such a dynamic environment and status of the SU call progress can be noticed till the end of requested holding time. The mean residual lifetime (MRL) of the free channels was computed based on requesting SU call holding time for PU channel allocation to a requesting SU. The channel with highest MRL is allocated. If the threshold probability in licensed band is taken to be 0.8, the success rate for CR users is achieved if the channels are ≥36 and holding time ≤ 2.5 min demanded by the SU. When the threshold probability for SU is 0.5, the success rate is achieved if the channels are ≥22 with minimum holding time ≤ 2.5 min demanded by the SU. Thus, success rate of channel allocation increases with the increase in number of channels in the system.

The proposed model shall be deployed to provide services for IoT through wireless communication. The implementation of the proposed work shall be a good solution where high volume of devices with low mobility is required for new wireless technologies.

Author details

Neeta Nathani[1]* and G. C. Manna[2]

*Address all correspondence to: neeta_nathani@yahoo.com

1 Electronics Engineering, G.H. Raisoni College of Engineering, Nagpur, Maharashtra, India

2 BSNL, Jabalpur, Madhya Pradesh, India

References

[1] Cisco Visual Networking Index: Forecast and Methodology, 2015–2020. February 3, 2016

[2] Ghosh G, Das P, Chatterjee S. Cognitive radio and dynamic spectrum access: A study. International Journal of Next-Generation Networks. March 2014;**6**(1):43-60. DOI: 10.5121/ijngn.2014.6104

[3] Mitola J, Maguire GQ Jr. Cognitive radio: Making software radios more personal. IEEE Personal Communications. August 1999;**6**(4):13-18

[4] Haykin S. Cognitive radio: Brain-empowered wireless communications. IEEE Journal on Selected Areas in Communications. February 2005;**23**(2):201-220. ISSN: 0733-8716

[5] Hiremath S. Transmission rate prediction for cognitive radio using adaptive neural fuzzy inference system. In: Proceedings of 5th International Conference on Industrial and Information Systems (ICIIS); July 29–August 1; September 2010. ISSN: 2164-7011

[6] Haykin S. Fundamental issues in cognitive radio. In: Cognitive Wireless Communication Networks. Boston, MA: Springer; 2007. pp. 1-43. DOI: https://doi.org/10.1007/978-0-387-68832-9. Print ISBN: 978- 0-387-68830-5

[7] Nathani N, Manna GC. Analysis of quality of service of cognitive radio systems. In: 2013 6th IEEE International Conference on Emerging Trends in Engineering and Technology (ICETET 13); IEEE Computer Society; 16-18 December 2013. pp. 107-108. ISBN: 978-1-4799-2561-2/13

[8] Nathani N, Manna GC. Quality of Service challenges for mobile cognitive radio. In: 2015 International Conference on Computer Science and Environmental Engineering; 17-18 May 2015; Beijing; China. DESTech Publications. pp. 967-998. ISBN: 978-1-60595-240-6

[9] Yucek T, Arslan H. A survey of spectrum sensing algorithms for cognitive radio applications. IEEE Communications Surveys Tutorials, First Quarter. 2009;**11**:116-130

[10] Al haj Mustafa SI, Amin Babiker A, Mustafa N. Comparison of detection techniques in spectrum sensing. International Journal of Science and Research (IJSR). May 2015;**4**(5): 1949-1953. ISSN (Online): 2319-7064

[11] Zahedi-Ghasabeh AT, Daneshrad B. Spectrum sensing of OFDM waveforms using embedded pilots in the presence of impairments. IEEE Transactions on Vehicular Technology. March 2012;**61**:1208-1221

[12] Bouzegzi A, Ciblat P, Jallon P. Matched filter based algorithm for blind recognition of OFDM systems. In: Proceedings: IEEE Vehicular Technology; September 2008. pp. 1-5

[13] Al-Habashna A, Dobre O, Venkatesan R, Popescu D. Cyclostationarity-based detection of LTE OFDM signals for cognitive radio systems. In: Proceedings: IEEE GLOBECOM 2010; December 2010. pp. 1-6

[14] Mourad Mabrook M, Hussein AI. Major spectrum sensing techniques for cognitive radio networks: A survey. International Journal of Engineering and Innovative Technology (IJEIT). September 2015;**5**(3):24-33. ISSN: 2277-3754

[15] Urkowitz H. Energy detection of unknown deterministic signals. In Proceedings: IEEE. April 1967;**55**(04):523-531a

[16] Ghasemi A, Sousa ES. Optimization of spectrum sensing for opportunistic spectrum access in cognitive radio networks. In Proceedings: IEEE CCNC. January 2007:1022-1026

[17] Bhargavi D, Murthy C. Performance comparison of energy, matched-filter and cyclostationarity-based spectrum sensing. In: Proceedings: IEEE SPAWC 2010; June 2010. pp. 1-5

[18] Tian Z, Giannakis G. Compressed sensing for wideband cognitive radios. In: Proceedings: IEEE ICASSP 2007, Vol. 4; April 2007. pp. 1357-1360

[19] Furtado A, Lúıs M, Oliveira R, Dinis R, Bernardo L. Channel availability assessment for cognitive radios. In: 4th Doctoral Conference on Computing, Electrical and Industrial Systems (DoCEIS); April 2013. pp. 495-504

[20] Moghaddam SS, Kamarzarrin M. A comparative study on the two popular cognitive radio spectrum sensing methods: Matched filter versus energy detector. American Journal of Mobile Systems, Applications and Services. 2015;**1**(2):132-139. Available from: http://www.aiscience.org/journal/ajmsas

[21] Federal Communications Commission. Spectrum Policy Task Force. Report of the Spectrum Efficiency Working Group; November 15; 2002

[22] Benitez ML, Umbert A, Casadevall F. Evaluation of spectrum occupancy in Spain for cognitive radio application. In Proceedings: IEEE 69th Conference on Vehicular Technology; VTC Spring 2009; Barcelona; April 26-29, 2009. pp. 1-5

[23] Islam Md H, Koh CL, Oh SW, Qing X, Lai YY, Wang C, Liang T, Toh BE, Chin F, Tan GL, Toh W. Spectrum survey in Singapore: Occupancy measurements and analysis. In: 3rd International Conference on Cognitive Radio Oriented Wireless Networks and Communications, 2008; CrownCom 2008, Singapore; May 15-17, 2008. pp. 1-7

[24] Erpek T, Lofquist M, Patton K. Spectrum Occupancy Measurements: Loring Commerce Center Limestone Maine; 18-20 September 2007

[25] Neeta N, Manna GC, Mule SB. An empirical assessment of quasi-permanently vacant channels in mobile communication bands for cognitive radio. ICACT Transactions on Advanced Communications Technology (TACT). January 2014;**3**(1):389-394

[26] Nathani N, Manna GC, Mule SB. Assessment of quasi-permanently vacant channels in mobile communication bands for cognitive radio. In: 15th IEEE International Conference on Advanced Communications Technology (ICACT 2013); 27-30 January 2013. pp. 484-488. ISBN: 978-89-968650-0-1

[27] Gao M-L. The self-adapt spectrum management formula base on cluster model. In: 2011 International Conference on Multimedia Technology (ICMT); 26-28 July 2011

[28] Parvin S, Hussain FK, Hussain OK, Han S, Tian B, Chang E. Cognitive radio network security: A survey. Journal of Network and Computer Applications. 2012;**35**(6):1691-1708

[29] Dere B, Bhujade S. An efficient spectrum decision making framework for cognitive radio networks. International Journal of Electrical and Electronics Engineers; IJEEE. Jan-June 2015;**07**(01):47-55. ISSN: 2321-2055 (E)

[30] Akyildiz IF, Lee W, Mehmet C, Mohanty S. A survey on spectrum management in cognitive radio networks. IEEE Communications Magazine; April. 2008;**46**(04):40-48. ISSN: 0163-6804

[31] Li H, Liu Y, Zhang D. Dynamic spectrum access for cognitive radio systems with repeated games. In: IEEE International Conference on Wireless Communications, Networking and Information Security (WCNIS); 25-27 June 2010

[32] Lee WY, Akyildiz IF. A spectrum decision framework for cognitive radio networks. IEEE Transactions on Mobile Computing. February 2011;**10**(02):1536-1233

[33] Shweta P. Spectrum sharing in cognitive radio using game theory. In: IEEE 3rd International Conference on Advance Computing Conference (IACC); 22-23 February 2013

[34] Grokop LH, David NC. Spectrum sharing between wireless networks. IEEE/ACM Transactions on Networking. October 2010;**18**(5):1401-1412

[35] Qun P, Zhang X, Chang Y, Yang D. Downlink dual modes multiple user MIMO cognitive radio system. In: 2nd IEEE International Conference on Network Infrastructure and Digital Content; 24-26 September 2010

[36] Nathani N, Manna GC. Call arrival rate prediction and blocking probability estimation for infrastructure based mobile cognitive radio personal area network. Advances in Science Technology and Engineering Systems Journal. August 2017;**2**(3):1609-1615. Special Issue on Recent Advances in Engineering Systems. ISSN: 2415-6698

[37] Nathani N, Manna GC, Dorle SS. Network architecture model of infrastructure based mobile cognitive radio system in licensed band with blocking probability assessment. In: IEEE Future Technologies Conference (FTC 2016); 6-7 December 2016. pp. 231-236. ISBN: 978-1-5090-4171-8/16

[38] Rao GS. Mobile Cellular Communication. India: Pearson Education; 2013. pp. 135-136. ISBN 978-81-317-7361-1

[39] Neeta N, Manna GC, Dorle SS. Channelized blocking probability estimation model for infrastructure based mobile cognitive radio. Indian Journal of Science and Technology. December 2016;**9**(45):1-7. DOI: 10.17485/ijst/2016/v9i45/106355, ISSN (Print): 0974-6846, ISSN (Online): 0974-5645

[40] Kalekar PS. Time Series Forecasting Using Holt-Winters Exponential Smoothing. 2004

[41] Introduction to time series analysis. Available from: http://www.itl.nist.gov/div898/handbook/pmc/section4/pmc4.htm

[42] Tirkeş G, Güray C, Çelebi N. Demand Forecasting: A Comparison between the Holt-Winters, Trend Analysis and Decomposition Models

[43] Nathani N, Manna GC, Mule SB. Dynamically available channel model for cognitive radio in GSM band. In Proceedings: National Conference on Trends in Signal Processing & Communication (TSPC'14); 12–14 April 2014. pp. 8-12. ISBN: 978-93-83842-40-7

[44] Hao C, Ljiljana T. Trunked radio systems: Traffic prediction based on user clusters. International Conference on Wireless Communication Systems. September 2004

[45] Li X, Seyed A, Zekavat R. Cognitive radio based spectrum sharing: Evaluating channel availability via traffic pattern prediction. Journal of Communications and Networks. 2009;**11**(2): 104-114

[46] Nathani N, Manna GC. Predicted call and residual lifetime based channel allocation model for PU equivalent QoS in cognitive radio cellular network. In: Second International Conference on Smart Trends for Information Technology and Computer Communications (SmartCom 2017); Conference Proceedings by Springer CCIS Series—Communications in Computer and Information Science; 18-19 August 2017. ISSN Number: 1865-0929

[47] http://www.statisticshowto.com/box-cox-transformation

Section 2

Spectrum Sensing and Spectrum Sharing in Cognitive Radio Systems

Chapter 3

Licensed Shared Access Evolution to Provide Exclusive and Dynamic Shared Spectrum Access for Novel 5G Use Cases

Juha Kalliovaara, Tero Jokela, Heikki Kokkinen and Jarkko Paavola

Additional information is available at the end of the chapter

http://dx.doi.org/10.5772/intechopen.79553

Abstract

This chapter studies the Licensed Shared Access (LSA) concept, which was initially developed to enable the use of the vacant spectrum resources in 2.3–2.4 GHz band for mobile broadband (MBB) through long-term static licenses. The LSA system was developed to guarantee LSA licensees a predictable quality of service (QoS) and exclusive access to shared spectrum resources. This chapter describes the development and architecture of LSA for 2.3–2.4 GHz band and compares the LSA briefly to the Spectrum Access System (SAS) concept developed in the USA. 5G and its new use cases require a more dynamic approach to access shared spectrum resources than the LSA system developed for 2.3–2.4 GHz band can provide. Thus, a concept called LSA evolution is currently under development. The novel concepts introduced in LSA evolution include spectrum sensing, short-term license periods, possibility to allocate spectrum locally, and support for co-primary sharing, which can guarantee the quality of service (QoS) from spectrum perspective. The chapter also describes a demonstration of LSA evolution system with spectrum user prioritization, which was created for Programme Making and Special Events (PMSE) use case.

Keywords: spectrum sharing, Licensed Shared Access (LSA), LSA evolution, PMSE, LTE, incumbents, 5G

1. Introduction

Demand for radio spectrum is constantly increasing as wireless services; especially, video streaming and emerging Internet of Things (IoT) are being adopted at an accelerating pace.

Mobile phones, laptops, and tablets are becoming more and more common, and the quality of available content and services is also increasing. This has resulted in rapid increases in the amount of traffic in mobile networks, and the increases are predicted to continue [1–3]. This presents extreme challenges for mobile communication systems, as there is a lack of new spectrum resources to be allocated for the growing number of connected devices, services, and users.

The wireless communication technologies themselves are approaching the fundamental theoretical limits of bandwidth efficiency, but simultaneously the frequency bands are exclusively licensed to different services which might not utilize all of their spectrum resources. Valuable spectrum resources can be left unexploited at different frequencies if the license owner does not use them at all times or at all locations. For example, several spectrum measurement campaigns covering frequencies up to 3 GHz state that the spectrum utilization rate is on the scale of 10–20% [4–6], and thus, most of the spectrum resources remain unused. It is necessary to utilize the existing frequency resources more efficiently to satisfy the growing demand for spectrum, but the current exclusive licensing methods do not allow this. Recent international studies have concluded that spectrum sharing will play a major role in maximizing the amount of available spectrum for wireless communication systems [7, 8].

The current exclusive spectrum licensing needs to be updated or replaced to enable spectrum sharing. In spectrum sharing, the users who currently hold an exclusive license to use a frequency band are called incumbents and are the primary users of the band. If the incumbents are using their spectrum resources inefficiently, their spectrum resources could potentially be shared with other users who could use the vacant spectrum resources at certain times or at certain locations where the license holder does not have any transmissions. Spectrum occupancy measurements have been proposed to find candidate frequency bands for spectrum sharing [4]. The vacant spectrum resources could be utilized through dynamic spectrum access methods, such as opportunistic spectrum access (OSA) [9] or Licensed Shared Access (LSA) [10]. In OSA, the shared spectrum user chooses the best available vacant transmission channel in an opportunistic and dynamic manner as an unlicensed secondary user of the spectrum, who does not need a license but does not have any guarantees on the amount and quality of available spectrum and has no protection from any harmful interference.

Cognitive radio spectrum sharing can be divided into three different types: underlay, overlay, and interweave. In underlay spectrum sharing, the cognitive users are allowed to operate if the interference they cause to the incumbents is below a given level. In overlay spectrum sharing, the cognitive user needs to know the incumbent signal. The cognitive user then adds its own data to the incumbent data and transmits the combined signal. In interweave spectrum sharing, the cognitive radios exploit spectral holes. The spectral holes are spectrum which is not used to be the incumbent in time, frequency, or spatial dimension. In each of the cognitive spectrum sharing types, accurate spectrum sensing data are of paramount importance both to guarantee the protection of the incumbents and to maximize the capacity available for the cognitive users. The currently standardized LSA belongs to interweave category, typically uses static vertical long-term spectrum leasing, and does not include spectrum sensing capabilities.

In LSA, vacant spectrum resources can be leased to shared spectrum users, known as LSA licensees, who are guaranteed an exclusive access to the leased spectrum resources and are protected from harmful interference. The incumbents are also protected from interference and might receive economic benefits from leasing their underutilized spectrum resources. The traffic load of the incumbent (licensed) users in LSA does not affect the performance of the LSA licensees, as the LSA licensees' transmissions are restricted so that they do not cause harmful interference to the incumbents under any circumstances. The terminology and definitions for shared spectrum access methods are diverse, but OSA and LSA could be considered as the two main categories in frequency bands with existing incumbents. Regardless of the used shared spectrum access method, it is essential to guarantee that the incumbents currently present in the band are protected from any harmful interference that could be induced by the newly introduced shared spectrum users.

The future LSA evolution will enable spectrum sensing and thus more dynamic use of spectrum. The current control solutions for network coordination are insufficient for heterogeneous 5G networks, where the performance of dense deployments could be further enhanced by advanced spectrum sharing [11]. 5G-PPP project called COHERENT considers the novel methods for coordinated control and spectrum management for 5G heterogeneous networks in LSA evolution.

Section 2 describes the development and architecture of LSA system for 2.3–2.4 GHz band. Section 3 discusses the feasibility, current status, and evolution of LSA toward 5G and makes a comparison to Spectrum Access System (SAS) concept developed in the USA. Section 4 describes an LSA evolution PMSE use case trial. Section 5 discusses the use of LSA evolution in 5G networks, and Section 6 gives the concluding remarks.

2. Development and architecture of LSA for 2.3–2.4 GHz band

The development of LSA concept began in European regulation and standardization to create a method for the mobile network operators (MNOs) to deploy their networks into bands allocated for mobile broadband (MBB), which currently have incumbents operating in the band. The concept allows spectrum sharing between an MNO and the incumbents with licensing conditions and rules that benefit both stakeholders. Radio Spectrum Policy Group (RSPG) proposed LSA concept [12] as an extension to an earlier proposal by an industry consortium, called Authorized Shared Access (ASA) [13]. ASA is limited to the International Mobile Telecommunications (IMT) use, while LSA can also be applied to other types of spectrum sharing. The 2.3 GHz band was chosen as the first frequency band for which to develop the operating conditions for LSA.

International Telecommunication Union Radiocommunication Sector (ITU-R) has globally allocated the 2.3–2.4 GHz band for mobile broadband (MBB) systems at the World Radiocommunication Conference 2007 (WRC-07) [14]. However, the frequency band is currently used by different incumbents in European Conference of Postal and Telecommunications Administrations (CEPT) countries [15]. The main users are PMSE applications, such as wireless camera links [16]. They are typically used to transmit video and audio wirelessly from a camera to an outside broadcasting (OB) van, and the typical users thus are the broadcasting companies.

The spectrum occupancy of the 2.3 GHz band in a single location in Finland was studied using several weeks of spectrum measurement data from Turku spectrum observatory in [17, 18]. The results showed that spectrum occupancy was very low and sporadic, and the detected busy periods were only 3–9 seconds long. The wireless camera transmissions typically occupy a bandwidth of 20 MHz, meaning a 20% occupancy per transmission over the whole 100 MHz frequency band. The instantaneous channel occupancy values were between 0 and 30%, but when the occupancy was filtered with a 5-minute moving average filter, the occupancy was between 0 and 5%. The filtered values confirm that the periods when the spectrum is occupied are very short in time. In addition to the signals interpreted as wireless cameras, only a small number of higher power peaks, probably from narrowband amateur radio services, was detected. The wireless camera transmissions are very low power and difficult to detect, and the studies conducted with a professional level wireless camera in [18] demonstrate that the spectrum observatories are able to detect the wireless cameras only from distances smaller than 250 m. Thus, single-location spectrum occupancy measurements cannot be used to draw strong conclusions on the spectrum occupancy trends over large geographical areas.

One reason why allocating the 2.3 GHz band for MBB in Europe is important is that the frequency band is already in MBB use in other regions. Thus, the transmitter hardware already exists and can be easily implemented in mobile receivers for European market. An economic analysis [19] also indicates that the impact of making 2.3 GHz band available for MBB in Europe could be worth 6.5–22 billion euros. However, the national administrations are unwilling to move the current incumbents to other frequency bands. Such an operation would result in expenses to the incumbents who would need to update their equipment, and in addition, there is a lack of suitable unallocated frequency bands. As the utilization of the 2.3 GHz frequency band appears to be very low, an optimal solution would be to let the current incumbents stay in the frequency band and to allow the MBB operation by exploiting the vacant spectrum resources. Again, the protection of the current incumbents is essential. LSA is needed in the 2.3 GHz band to provide exclusive shared spectrum access to the MBB and to protect the current incumbents.

Working Group Frequency Management (WG FM) established Frequency Management 53 (FM53)—Reconfigurable Radio Systems (RRS) and LSA project team in September 2012. The aim of FM53 was to provide generic guidelines to CEPT administrations for the implementation of the LSA. The European Commission (EC) requested an opinion from RSPG on regulatory and economic aspects of LSA in November 2012 [20], and their final opinion from November 2013 [21] defined that LSA is "a regulatory approach aiming to facilitate the introduction of radiocommunication systems operated by a limited number of licensees under an individual licensing regime in a frequency band already assigned or expected to be assigned to one or more incumbent users. Under the LSA approach, the additional users are authorized to use the spectrum (or part of the spectrum) in accordance with sharing rules included in their rights of the use of spectrum, thereby allowing all the authorized users, including incumbents, to provide a certain QoS."

Thus, LSA gives the MNOs a predictable QoS through individual licensing and exclusive shared access to the spectrum resources. The MNO accessing shared spectrum through

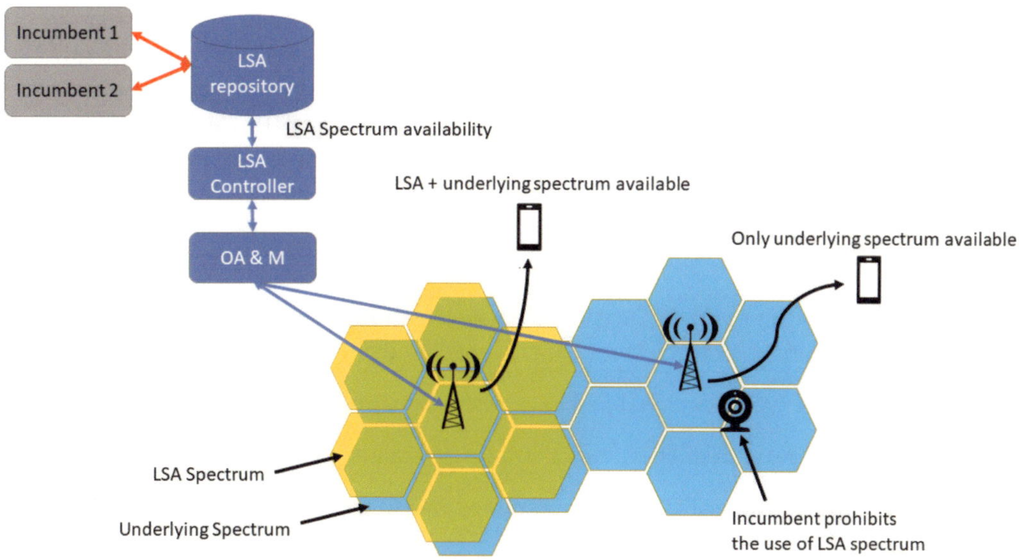

Figure 1. LSA architecture (Adapted from [22]).

temporary leasing is called LSA licensee. The functionalities of LSA are enabled mainly by two additional units on top of the existing mobile networks: the LSA Repository and the LSA Controller. The LSA Repository is a database containing information on incumbent spectrum utilization, while the task of the LSA Controller is to guarantee protection and interference-free operation for both types of users by using the data from the LSA Repository. The LSA Repository can be managed by the National Regulatory Authorities, the incumbents, or a trusted third party.

The LSA Repository contains information on the spectrum availability for LSA licensees and spectrum sharing rules. This information is communicated to the LSA Controller through a secure and reliable communication path. Based on the information from the LSA Repository, the LSA Controller controls the spectrum use of LSA licensee(s). There may be several LSA Repositories from which the LSA Controller gets the information on spectrum availability and also several LSA licensees' networks.

Figure 1 illustrates the LSA architecture. Several incumbents provide information on their spectrum utilization to the LSA Repository, which communicates it to the LSA Controller. The LSA Controller provides this information to the MNO operations, administration, and maintenance (OAM), which instructs that the relevant base stations of the MBB network can use the spectrum resources which are not used by the incumbents in the band. These newly available spectrum resources are taken into use to provide additional capacity through carrier aggregation (CA). The underlying spectrum in other frequency bands (blue cells in the figure) is exclusively licensed for MBB transmissions, while the orange cells can provide additional capacity using the LSA spectrum resources in the 2.3 GHz band. On the right side of the figure, the incumbent operation prevents the use of LSA spectrum, and only the underlying MBB spectrum resources can be used. This is illustrated through the absence of orange LSA cells.

LSA Spectrum Resource Availability Information (LSRAI) contains the information on the LSA spectrum resource that may be used by the LSA licensee. LSRAI is generated in the LSA Repository and sent to the LSA Controller over the LSA_1 interface using LSA_1 protocol messages as defined in the ETSI technical specification [23]. The LSA Information Exchange Function to maintain the LSRAI synchronization between LSA Controllers and LSA Repositories and the LSA_1 protocol are described in detail in [23].

As defined in [23], LSRAI has the following characteristics:

- It contains one or more zones. A zone is an information object which describes a set of operational conditions or restrictions to be applied by the LSA licensee.
- A zone has a zone type associated to it (e.g., restriction, protection, exclusion).
- A zone contains space, frequency, radio, and time parameters:
    - Space parameters describing the geographical area to which the restriction applies.
    - Frequency parameters describing the frequency range to which the restriction applies.
    - Time parameters describing when the restriction applies.
    - Radio parameters describing the RF restrictions to be applied within the space/frequency/time combination defined by the above parameters
- A zone has a zone ID and a zone configuration index associated to it.

3. The status of LSA and SAS

This section considers the feasibility, current status, and evolution of LSA and briefly compares it to the US concept for licensed shared spectrum access; Spectrum Access System (SAS). The work on LSA has been very active in regulation and standardization: CEPT Reports [24–26], ECC harmonized conditions for the use of the 2.3 GHz band in [15, 27–29], and European Telecommunications Standards Institute (ETSI) standardization in [22, 23, 30, 31] provide all the measures needed for a National Regulatory Authorities in a CEPT country to create an implementation of LSA. A regulatory evaluation in [32] concluded that LSA implementations are feasible as they provide a simple spectrum sharing approach providing a high degree of certainty for both the incumbent and the LSA licensee with low impact to the systems and the concept has already been tested and approved. The use of LSA is a national matter, which does not require modifications to the ITU-R Radio Regulations (RR) but needs to comply with the current regulations.

A study on the feasibility of LSA from business perspective [33] concluded that LSA implementations could be profitable for MNOs in Finland if they have a reasonably good customer base and well-defined network launch and management. Most importantly, the MNO has to carefully investigate the techno-economics to see if there is a customer base large enough to justify the investments in new spectrum and network resources. A Finnish LSA trial environment is operated in Ylivieska [34, 35], but no commercial deployments of LSA in 2.3 GHz band are available yet. A service pilot with LSA radio licenses to commercial end users operating with incumbent wireless cameras in the 2.3 GHz band was announced in the Netherlands in May 2016 [36], and more pilots are expected in the near future. The LTE MNOs are expected to make fairly static multiyear spectrum sharing contracts with the incumbents to justify investments in building mobile network infrastructure for LSA operation [37]. LSA could also provide mechanisms to mitigate intra-MNO-system interference [37].

A concept called SAS is in development in the USA. It is very similar to LSA, as both of them include incumbent users and licensed shared users who have exclusive shared access to the spectrum. The Licensed Shared Access in SAS is known as Priority Access License (PAL). LSA excludes opportunistic access where no protection from incumbents is provided, but SAS adds an additional third tier for unlicensed opportunistic spectrum access with General Authorized Access (GAA), as shown in **Figure 2**. PAL users are protected from interference from GAA tier, but not from the incumbents.

The SAS design ensures protection also for the incumbents who cannot provide a priori information to a central database. This is a major difference to LSA, where this information has to be communicated to a central database (LSA Repository) in order to protect the incumbents. The incumbents operating in the Citizens Broadband Radio Service (CBRS) band include military services whose information is too sensitive to be stored in a database. Instead, SAS includes Environmental Sensing Capability (ESC) component which uses spectrum sensing to provide the needed data for spectrum access decisions. As [38] states, spectrum sensing is not a trivial matter, especially with the strict requirements in SAS. ESC will not be used in the first phase of SAS deployment, which restricts the SAS operation in the zones with military incumbents near coastal areas until a suitable ESC technology is available. ESC technologies have

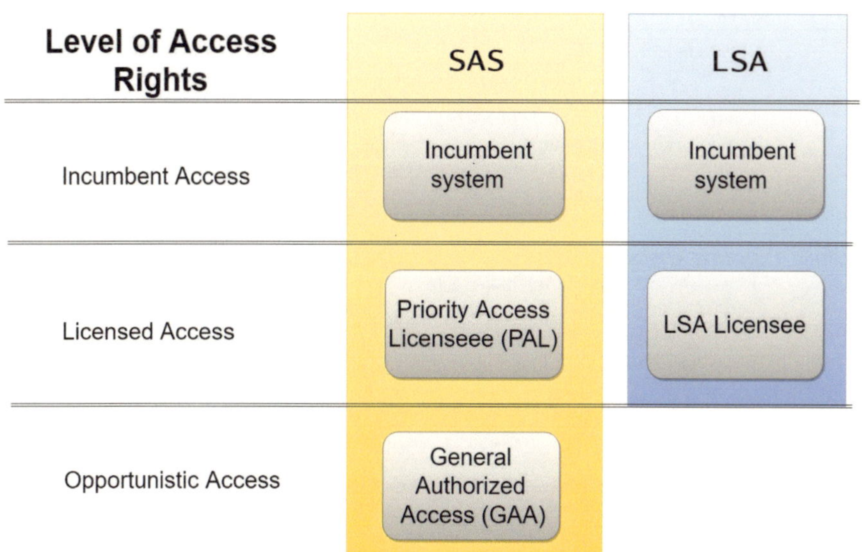

Figure 2. Overview of the level of access rights in different tiers of SAS and LSA sharing models.

already been developed and demonstrated in SAS trials [39]. Unlike LSA, SAS standardization is still in progress, but the industrial interest in CBRS Alliance [40] is strong and advances are expected in the near future. The first commercial SAS deployments are expected during 2018 [41] in 3.55–3.7 GHz CBRS [42] band in the USA. **Table 1** gives a brief comparison of the key features of the LSA and SAS concepts. More detailed considerations and comparisons from both technical and business perspectives are given in [43, 44].

LSA and SAS are currently defined for use only in the mentioned frequency bands with their specific incumbents, but the basic operational principles are straightforward to adopt to other sub-6 GHz frequency bands, where spectrum sharing is more relevant. Spectrum sharing is less relevant in higher frequency ranges, such as mmWaves, where wireless communication is not so much limited by interference, but the higher path losses. The ETSI LSA standardization was done partly in liaison with the 3rd Generation Partnership Project (3GPP) [45], which has studied how LSA could provide a global solution for a 3GPP MNO in [46]. LSA has also been recognized as one of the future technology trends for IMT in the ITU-R Working Party 5D on IMT systems [47].

5G and its new use cases require a more dynamic approach to access shared spectrum resources than ETSI LSA for 2.3–2.4 GHz band can provide. Spectrum sensing techniques are needed as the more dynamic access to spectrum cannot be achieved by using static a priori information. The dominant problems in spectrum sensing are the removal of shadowing and multipath fading. Methods to overcome these problems through cooperative mobile measurements to create interference maps are discussed in [48], but the current technologies related to spectrum sensing are still not able to guarantee protection from harmful interference [43].

	LSA	SAS
Tiers	Two tiers with individual access	3-tier system; two tiers with individual access and a third license-exempt tier
Database	Centralized geo-location database based on static a priori information on the incumbents	Centralized geolocation database with information based on spectrum sensing technologies
Spectral efficiency	Less efficient use of spectrum	More efficient use of spectrum
Use of spectrum	Current version is a static framework for long-term spectrum leasing in 2.3 GHz band. Future LSA evolution will include spectrum sensing to provide more dynamic use of spectrum	GAA tier enables very flexible and dynamic short-term use of spectrum with a very low entry barrier, but the GAA spectrum access and quality are less certain than with the exclusive LSA licenses
Software and hardware	Minimal additions to the existing 3GPP network ecosystem.	Requires new near real-time sensing capabilities and big data and spectrum analytics
Complexity	Less complex	Very complex due to the spectrum sensing needed for the GAA tier
Adaptability	Initially focused on Europe but easily adaptable to other regions	Initially specific to US federal use, additional adaptability is needed for other regions

Table 1. Brief comparison of LSA and SAS concepts.

A concept called LSA evolution is currently under development. The use of spectrum sensing is considered to provide more dynamic version of LSA, which is needed for the novel 5G use cases [49, 50]. The original LSA specification assumes that the spectrum is available for the operator always when the incumbent does not use the spectrum. The spectrum is available for the operator within the regulative area, like country borders, excluding the areas where the incumbent uses the spectrum. From spectrum perspective, high QoS is achieved when the incumbent does not use the spectrum.

The interest in private LTE and 5G networks has grown due to the increased number of IMT frequency bands, higher frequency ranges, variety in spectrum assignments for 3GPP technologies, and revolution of wireless industrial communication [51]. The feasibility study [52] addresses these issues and applies learning from the later developed SAS/CBRS system at the same time. The study considers, for example, how to provide temporary spectrum access for local high-quality wireless networks.

The new concepts for LSA evolution include short-term license periods, possibility to allocate spectrum locally, and supporting co-primary sharing, which can guarantee the quality of service from spectrum perspective [49, 53]. Most LTE use is static, when the spectrum assignments are considered. Even if the user equipments (UEs) are mobile, the spectrum use is more determined by the eNodeBs. They traditionally require masts, electricity, backhaul connectivity, and professional installers.

The temporary LTE or 5G spectrum access is most likely to be related to PMSE, Public Protection and Disaster Relief (PPDR), or Test and Development (T&D) licenses. The current mobile networks are wide area networks even if they are built for capacity. Most private LTE and 5G networks are local. PMSE, PPDR, and T&D networks are both temporary and local and thus can benefit most from LSA evolution.

4. LSA evolution systvem for 5G PMSE use case trial

This section presents a trial of LSA evolution system developed for 5G PMSE use case. The trial focuses on utilizing LSA for sharing spectrum in 2.3–2.4 GHz band between wireless cameras (PMSE) and mobile network operator (MNO) serving users. When the spectrum is required by the incumbents, such as wireless video cameras during a sports event, the transmissions of the mobile network in this area need to be controlled to allow the operation of the wireless cameras in the band. The mobile network base stations on this band can be shut down or their transmission powers and potentially operating frequencies controlled.

The developed LSA evolution system allows to set priorities for different users of the spectrum, and thus it is possible, for example, to give the highest priority to the old/proprietary PMSE systems which cannot communicate bidirectionally with the spectrum manager, which includes the functionalities of LSA Controller and LSA Repository. The LTE/5G-based equipment can be controlled (their transmission frequencies and power levels adjusted) so that no interference is caused toward the old/proprietary PMSE equipment or other LTE/5G-based equipment.

The trial assumes that the broadcasters and other PMSE stakeholders may have a mixture of proprietary and LTE/5G PMSE wireless technology in use in the future. This trial demonstrates how broadcasters can gradually move from proprietary 2.3 GHz wireless camera technology to 2.3 GHz LTE/5G PMSE. Both old and new equipment can be used simultaneously within the trial system. One major advantage of LTE/5G radio-based PMSE is that the spectrum manager can directly control the equipment (e.g., change its operating frequency to avoid interference). Another advantage of having an own PMSE LTE/5G system compared to using commercial LTE/5G networks for the PMSE traffic is that the PMSE stakeholder is able to fully control the use and thus the load of its own PMSE system.

The architecture of the trial setup shown in **Figure 3** consists of PMSE equipment operating occasionally on 2.3 GHz band and MNO LTE network operating on 700 and 2.3 GHz bands. This represents a situation where MNO employs additional capacity on 2.3 GHz band using, for example, supplemental downlink (SDL) concept. Proprietary PMSE equipment represents an OFDM-based proprietary solution for wireless cameras operating on the band. PMSE LTE in **Figure 3** is a rapidly deployable LTE/5G network for PMSE purposes. Commercial base stations and LTE terminals were used in the trial. The proprietary PMSE equipment was emulated in the trial with a DVB-T/DVB-T2 transmission and Samsung S8 phones streaming video served as LTE-based PMSE equipment.

Spectrum manager orchestrates the operation of the different systems on 2.3 GHz shared band. PMSE system information is collected with a web-based reservation system, where the users of the devices can make reservations for their intended use. The reservation system has been piloted in the Netherlands in 2017–2018 [34]. The control of the PMSE devices also takes place through the reservation system so that the user of the devices is informed about the required spectrum use changes and the user has to deploy the configuration changes in their devices.

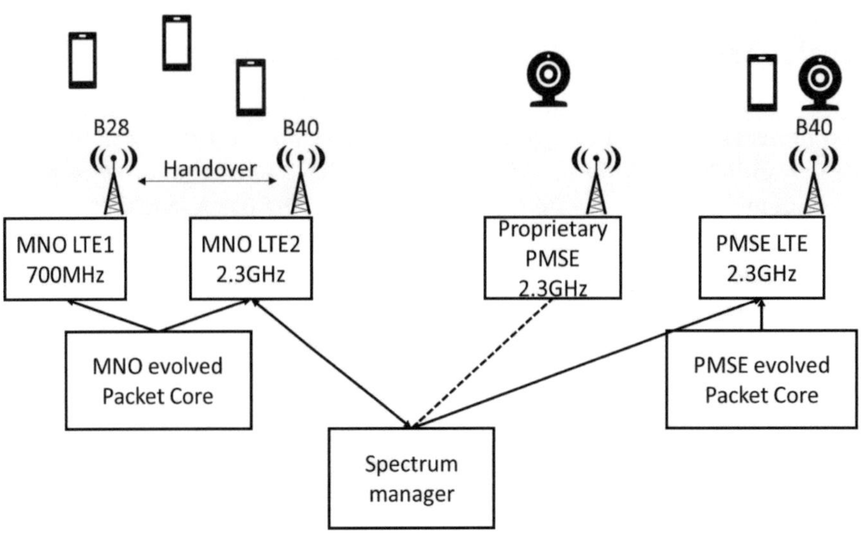

Figure 3. Spectrum demonstration architecture.

Both PMSE LTE and MNO LTE systems have a direct machine-to-machine (M2M) interface between the radio equipment and the spectrum manager. The priority order considered in the trial is as follows, from highest to lowest: PMSE, PMSE LTE, and MNO LTE. When the priority user changes the configuration of the LTE network, a notification about the change is automatically received in the spectrum manager. The spectrum manager processes the changed spectrum situation and evaluates if the lower priority use may cause harmful interference to the higher priority use. If there is a risk of interference, the spectrum manager evaluates which changes would be required to accommodate the higher priority use and to maintain the best possible service level also for the lower priority use.

On the high level, interference mitigation is implemented so that if there are frequency channels available, the lower priority use is transferred to those channels. If there are no other channels available, the power level of the secondary user is reduced or the transmission is denied. In this demonstration, the higher priority user is able to select the frequency channel to be used. An option for this could be that the higher priority user has the right to the spectrum resource in the band, but the specific frequency channel is determined by the spectrum manager.

The main target of the performed trial is to demonstrate the LSA evolution functions developed to the spectrum manager to enable dynamic spectrum sharing between users with different levels of priority. The steps performed in the trial were:

1. MNO LTE1 (700 MHz) and MNO LTE2 (2.3 GHz) serving users (web surfing, video streaming).
2. PMSE LTE (2.3 GHz) turns on as a rapidly deployable network for PMSE, and spectrum is available for both MNO LTE2 and PMSE LTE.
3. PMSE user registers to the spectrum manager registration system, on the frequency currently in use for PMSE LTE.
4. MNO LTE2 limits its transmission power (if necessary) to follow interference limits, and the users remain connected to at least B28 (700 MHz) base station.
5. PMSE LTE changes channel to give space to PMSE.
6. Proprietary PMSE equipment turns on.

Corresponding snapshots of the 2.3 GHz spectrum band are illustrated in **Figure 4**. First, the lowest priority LTE service, such as SDL of MNO LTE2, operates in the band. Then, PMSE using rapidly deployable LTE air interface (PMSE LTE) requests for spectrum. At the same time, there is enough free spectrum for both to operate. Then, the proprietary PMSE equipment requests for spectrum, and the spectrum manager allocates suitable frequencies and power levels for all users. If necessary, MNO LTE2 adjusts the transmission power according to regulated interference limits to allow for the operation of higher priority users. Also, PMSE LTE that is controlled by the spectrum manager via M2M interface switches frequency (e.g., due to the limitations of proprietary PMSE equipment tuning range). Finally, all three networks operate on the shared spectrum without causing interference to each other.

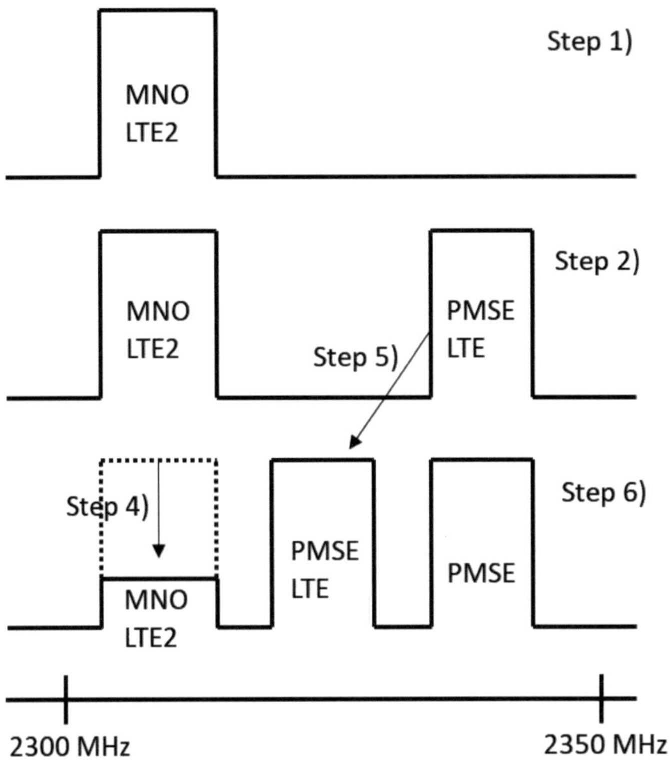

Figure 4. 2.3 GHz spectrum use corresponding to trial steps.

5. LSA evolution in 5G networks

A mobile service of a mobile network operator consists of different mobile network technologies, like GSM, WCDMA, LTE, and 5G. Each of the technologies, especially LTE, has several frequency bands. The bands below 1 GHz are coverage bands, and the bands above 2 GHz are capacity bands. The capacity bands are available only in densely populated areas.

Mobile phones primarily make the decision which technology and which band(s) they use. The availability of the LSA secondary bands cannot be guaranteed at any time or location, but the situation does not differ much from the availability of the capacity bands, when considering the availability of the bands from the mobile device perspective. At an arbitrary location and time, only a part of the deployed technologies and frequency bands are available for a specific mobile device. The generic secondary LSA spectrum use fits best to 5G enhanced Mobile Broadband (eMBB).

The original LSA was developed to allow mobile as a secondary user on the bands, which have other types of priority users. The recent development in ETSI RRS considers LSA for local high-quality networks. The main issue to ensure a guarantee of quality is to have a sharing agreement, where the LSA user is the primary user and is protected from interference.

When the LSA users have a primary status and when they are protected from interference, LSA can be used also for 5G Ultrareliable Low-Latency Communication (URLLC).

When the 5G massive Machine-Type Communication (mMTC) is deployed in coverage networks, LSA may not be the best solution, as the spectrum sharing in the coverage bands is not as beneficial as in the capacity bands. On the other hand, many of the sub-GHz wide-area IoT networks operate on license-exempt bands, which cannot guarantee quality of service either.

The LSA spectrum sharing does not change mobility or handovers in the mobile networks compared to non-LSA mobile networks. The main issue in this respect is the graceful shutdown. The sharing agreement may allow a reasonable delay, i.e., several minutes or more, between the moment information that the primary user requiring interference protection arrives and the moment when the interference protection has to be carried out in the LSA system. In that case, the operations, administration, and maintenance (OAM) of the mobile network can force the mobile to non-LSA bands before the LSA system deploys the interference protection in the LSA band. The graceful shutdown is not a part of the LSA system but rather a feature of the OAM.

6. Conclusion

This chapter has discussed why spectrum sharing is needed and introduced the LSA concept developed to provide a predictable QoS and exclusive access to shared spectrum resources. The first phase of LSA development and standardization created a somewhat static system and rules for the use of LSA in the 2.3–2.4 GHz frequency band. This version of LSA is best suited to facilitate access to sub-6 GHz frequency bands where the existing incumbents are not efficiently using their spectral resources.

However, the novel use cases in 5G require a more dynamic access to the spectrum and novel solutions for coordinated control and spectrum management. Spectrum sensing techniques are needed to provide the more dynamic access to spectrum, as the current version of LSA and its static spectrum allocations are insufficient for this. The spectrum sensing techniques however still need to evolve to be able to guarantee protection from harmful interference.

The development of LSA evolution is underway, and the other new concepts needed for LSA evolution include short-term license periods, possibility to allocate spectrum locally, and support for co-primary sharing, which can guarantee the QoS from spectrum perspective. The chapter also described a demonstration of LSA evolution system with spectrum user prioritization, which was created for 5G PMSE use case.

Acknowledgements

This work is supported by Business Finland under the project Critical Operations over Regular Networks (CORNET). The views expressed in this contribution are those of the authors and do not necessarily represent the projects.

Author details

Juha Kalliovaara[1], Tero Jokela[1], Heikki Kokkinen[2] and Jarkko Paavola[1]*

*Address all correspondence to: jarkko.paavola@turkuamk.fi

1 Turku University of Applied Sciences, Turku, Finland

2 Fairspectrum, Espoo, Finland

References

[1] Cisco. Cisco visual networking index: Global mobile data traffic forecast update, 2016-2021 white paper. Feb 2017

[2] Ericsson. Ericsson mobility report 2017. Nov 2017

[3] ITU-R. Report ITU-R M.2290: Future spectrum requirements estimate for terrestrial IMT. Dec 2013

[4] McHenry MA, Tenhula PA, McCloskey D, Roberson DA, Hood CS. Chicago spectrum occupancy measurements & analysis and a long-term studies proposal. In: Proceedings of the First International Workshop on Technology and Policy for Accessing Spectrum (TAPAS 06). 2006

[5] Valenta V, Fedra Z, Marsalek R, Baudoin G, Villegas M. Towards cognitive radio networks: Spectrum utilization measurements in suburb environment. In: 2009 IEEE Radio and Wireless Symposium. Jan 2009. pp. 352-355

[6] Lopez-Benitez M, Umbert A, Casadevall F. Evaluation of spectrum occupancy in Spain for cognitive radio applications. In: Vehicular Technology Conference; 2009. VTC Spring 2009. IEEE 69th; Apr 2009. pp. 1-5

[7] European Commission. COM/2012/478: Promoting the shared use of radio spectrum resources in the internal market. Sept 2012

[8] President's Council of Advisors on Science and Technology. Realizing the full potential of government-held spectrum to spur economic growth. Jul 2012

[9] Zhao Q, Sadler BM. A survey of dynamic spectrum access. IEEE Signal Processing Magazine. May 2007;**24**(3):79-89. DOI: 10.1109/MSP.2007.361604

[10] Ponomarenko-Timofeev A, Pyattaev A, Andreev S, Koucheryavy Y, Mueck M, Karls I. Highly dynamic spectrum management within licensed shared access regulatory framework. IEEE Communications Magazine. Mar 2016;**54**(3):100-109. DOI: 10.1109/MCOM.2016.7432155

[11] COHERENT. Deliverable D4.1: Report on enhanced LSA, intra-operator spectrum-sharing and micro-area spectrum sharing, Coordinated control and spectrum management

for 5G heterogeneous radio access networks [Online]. 2015. Available from: http://www.ict-coherent.eu/coherent/wp-content/uploads/2015/10/COHERENT_D4_1_v1.pdf

[12] RSPG. RSPG11-392: Report on Collective Use of Spectrum (CUS) and other spectrum sharing approaches. Nov 2011

[13] CEPT/ECC. FM(12)084 Annex 47: Report on ASA concept. 2012

[14] ITU. Final acts WRC-07: World Radiocommunication Conference (Geneva 2007). 2007

[15] ECC. ECC decision (14)02: Harmonised technical and regulatory conditions for the use of the band 2300-2400 MHz for Mobile/Fixed Communications Networks (MFCN). Jun 2014

[16] ECC. ECC document FM(12)017rev1: Results of the WG FM QUESTIONNAIRE to CEPT ADMINISTRATIONS on the current and future usage of frequency band 2300-2400 MHz. Apr 2012

[17] Höyhtyä M, Matinmikko M, Chen X, Hallio J, Auranen J, Ekman R, Röning J, Engelberg J, Kalliovaara J, Taher T, Riaz A, Roberson D. Measurements and analysis of spectrum occupancy in the 2.3-2.4 GHz band in Finland and Chicago. In: 9th International Conference on Cognitive Radio Oriented Wireless Networks and Communications (CROWNCOM), Jun 2014. 2014. pp. 95-101

[18] Höyhtyä M, Matinmikko M, Chen X, Hallio J, Auranen J, Ekman R, Röning J, Engelberg J, Kalliovaara J, Taher T, Riaz A, Roberson D. Spectrum occupancy measurements in the 2.3-2.4 GHz band: Guidelines for licensed shared access in Finland. EAI Endorsed Transactions on Cognitive Communications. May 2015;**1**(2),e2:1-11. DOI: 10.4108/cogcom.1.2.e2

[19] Lavender T, Marks P, Wongsaroj S. The economic benefits of LSA in 2.3 GHz in Europe. Dec 2013

[20] RSPG. RSPG12-424 final: RSPG request for opinion on licensed shared access (LSA). Nov 2012

[21] RSPG. RSPG13-538: RSPG opinion on licensed shared access. Nov 2013

[22] ETSI. ETSI TR 103 113 v1.1.1.: Mobile broadband services in the 2300-2400 MHz frequency band under Licensed Shared Access regime. Jul 2013

[23] ETSI. ETSI TS 103 379 V1.1.1: Reconfigurable radio systems (RRS); Information elements and protocols for the interface between LSA controller (LC) and LSA repository (LR) for operation of licensed shared access (LSA) in the 2300 MHz–2400 MHz band. Jan 2017

[24] CEPT. CEPT report 55: Report A from CEPT to the European commission in response to the mandate on harmonised technical conditions for the 2300-2400 MHz (2.3 GHz) frequency band in the EU for the provision of wireless broadband electronic communications services: Technical conditions for wireless broadband usage of the 2300-2400 MHz frequency band. Nov 2014

[25] CEPT. CEPT report 56: Report B1 from CEPT to the European commission in response to the mandate on harmonised technical conditions for the 2300-2400 MHz (2.3 GHz) frequency band in the EU for the provision of wireless broadband electronic communications services: Technological and regulatory options facilitating sharing between wireless broadband applications (WBB) and the relevant incumbent service/application in the 2.3 GHz band. Mar 2015

[26] CEPT. CEPT report 58: Report B2 from CEPT to the European commission in response to the mandate on harmonised technical conditions for the 2300-2400 MHz (2.3 GHz) frequency band in the EU for the provision of wireless broadband electronic communications services: Technical sharing solutions for the shared use of the 2300-2400 MHz band for WBB and PMSE. Jul 2015

[27] ECC. ECC report 205: Licensed shared access (LSA). Feb 2014

[28] ECC. ECC recommendation (14)04: Cross-border coordination for mobile/fixed communications networks (MFCN) and between MFCN and other systems in the frequency band 2300-2400 MHz. May 2014

[29] ECC. ECC recommendation(15)04: Guidance for the implementation of a sharing framework between MFCN and PMSE within 2300-2400 MHz. Jul 2015

[30] ETSI ETSI TS 103 154 V1.1.1: Reconfigurable radio systems (RRS); System requirements for operation of mobile broadband systems in the 2300 MHz–2400 MHz band under licensed shared access (LSA). Oct 2014

[31] ETSI. ETSI TS 103 235 V1.1.1: Reconfigurable radio systems (RRS); System architecture and high level procedures for operation of licensed shared access (LSA) in the 2300 MHz–2400 MHz band. Oct 2015

[32] Mustonen M, Matinmikko M, Roberson D, Yrjölä S. Evaluation of recent spectrum sharing models from the regulatory point of view. In: 1st International Conference on 5G for Ubiquitous Connectivity (5GU), Nov 2014. 2014. pp. 11-16

[33] Luttinen E, Matinmikko M, Ahokangas P, Katz M, Yrjölä S. Feasibility assessment of licensed shared access (LSA) concept—Case of a Finnish mobile network operator (MNO). In: 1st International Conference on 5G for Ubiquitous Connectivity (5GU), Nov 2014. 2014. pp. 252-257

[34] Palola M, Matinmikko M, Prokkola J, Mustonen M, Heikkilä M, Kippola T, Yrjölä S, Hartikainen V, Tudose L, Kivinen A, Paavola J, Heiska K. Live field trial of licensed shared access (LSA) concept using LTE network in 2.3 GHz band. In: IEEE International Symposium on Dynamic Spectrum Access Networks (DYSPAN), Apr 2014. 2014. pp. 38-47

[35] Matinmikko M, Palola M, Saarnisaari H, Heikkilä M, Prokkola J, Kippola T, Hänninen T, Jokinen M, Yrjölä S. Cognitive radio trial environment: First live authorized shared

access-based spectrum-sharing demonstration. Vehicular Technology Magazine, IEEE. Sept 2013;**8**(3):30-37. DOI: 10.1109/MVT.2013.2269033

[36] Radiocommunications Agency of the Netherlands. Licensed shared access pilot in the Netherlands [Online]. 2016. Available from: https://cept.org/Documents/wg-fm/33535/5-3 pmse-lsa-in-the-band-2300-2400-mhz-nl [Accessed: 20-04-2018]

[37] Badic B, Drewes C, Karls I, Mueck M. Rolling out 5G: Use Cases, Applications, and Technology Solutions. New York, US: Apress; 2016

[38] Vartiainen J, Höyhtyä M, Vuohtoniemi R, Ramani VL. The future of spectrum sensing. In: 2016 Eighth International Conference on Ubiquitous and Future Networks (ICUFN). Jul 2016. pp. 247-252

[39] Aho P, Palola M, Kippola T, Heikkilä M, Mäkeläinen M, Hänninen T, Tudose L, Hartikainen V, Yrjölä S, Kivinen A, Paavola J. Field trial of citizens broadband radio service (CBRS)/spectrum access system (SAS). In: Wireless Innovation Forum European Conference on Communications Technology and Software Defined Radio (WInnComm-Europe 2016); Paris, France. Oct 2016

[40] CBRS Alliance [Online]. Available from: http://www.cbrsalliance.org/ [Accessed: 25-04-2018]

[41] CBRS WInnForum Standards. CBRS Status Summary [Online]. Available from: https://www.cbrs.wirelessinnovation.org/cbrs-status-summary [Accessed: 25-04-2018]

[42] FCC. 3.5 GHz Band/Citizens Broadband Radio Service [Online]. Available from: https://www.fcc.gov/rulemaking/12-354 [Accessed: 25-04-2018]

[43] Massaro M. Next generation of radio spectrum management: Licensed shared access for 5G. Telecommunications Policy. Jun 2017;**41**(5-6):422-433. DOI: 10.1016/j.telpol.2017.04.003

[44] Yrjölä S, Matinmikko M, Mustonen M, Ahokangas P. Spectrum sharing transforms mobile broadband networks towards markets—Analysis of sharing economy antecedents for recent spectrum sharing concepts. International Journal on Advances in Intelligent Systems. Dec 2017;**10**(3&4):410-422

[45] Mustonen M, Matinmikko M, Palola M, Rautio T, Yrjölä S. Analysis of requirements from standardization for licensed shared access (LSA) system implementation. In: IEEE International Symposium on Dynamic Spectrum Access Networks (DySPAN), Sept 2015. 2015. pp. 71-81

[46] 3GPP. 3GPP TR 32.855 V14.0.0: Study on OAM support for licensed shared access (LSA); (Release 14). Oct 2015

[47] ITU-R. Report ITU-R M.2320-0: Future technology trends of terrestrial IMT systems. Nov 2014

[48] Höyhtyä M, Mammela A, Eskola M, Matinmikko M, Kalliovaara J, Ojaniemi J, Suutala J, Ekman R, Bacchus R, Roberson D. Spectrum occupancy measurements: A survey and use of interference maps. IEEE Communications Surveys & Tutorials;**18**(4):2386-2414, Fourthquarter 2016. DOI: 10.1109/COMST.2016.2559525

[49] Yrjölä S, Kokkinen H. Licensed shared access evolution enables early access to 5G spectrum and novel use cases. EAI Endorsed Transactions on Wireless Spectrum. Dec 2017;**3**(10),e1:1-11. DOI: 10.4108/eai.12-12-2017.153463

[50] Frascolla V, et al. Dynamic licensed shared access—A new architecture and spectrum allocation techniques. In: 2016 IEEE 84th Vehicular Technology Conference (VTC-Fall); Montreal, QC. 2016. pp. 1-5. DOI: 10.1109/VTCFall.2016.7880971

[51] Kokkinen H, Yrjölä S, Engelberg J, Kokkinen T. Pricing private LTE and 5G radio licenses on 3.5 GHz. Cognitive Radio Oriented Wireless Networks and Communications (CROWNCOM). 2018

[52] ETSI. ETSI TR 103 588 V1.1.1: Reconfigurable radio systems (RRS); Feasibility study on temporary spectrum access for local high-quality wireless networks. Feb 2018

[53] Guirao MDP, Wilzeck A, Schmidt A, Septinus K, Thein C. Locally and temporary shared spectrum as opportunity for vertical sectors in 5G. IEEE Network. Nov 2017;**31**(6):24-31. DOI: 10.1109/MNET.2017.1700079

Chapter 4

Spectrum Decision Framework to Support Cognitive Radio Based IoT in 5G

Ahmad Naeem Akhtar, Fahim Arif and
Adil Masood Siddique

Additional information is available at the end of the chapter

http://dx.doi.org/10.5772/intechopen.80991

Abstract

RF Spectrum Decision in Cognitive Radio enables unlicensed users of wireless communication systems to occupy the vacant spectrum slots as a solution to scarce spectrum. Internet of Things (IoT) is a wide-reaching network of unified entities. IoT capable things will be interconnected through wireless communication technologies offering cost-effectiveness and accessibility to remote users making quality life style. IoT implementation suffers from challenges of vulnerabilities to dynamic environmental conditions, ease of access, bandwidth allocation and utilization, and cost to purchase RF spectrum. As RF spectrum is a precious commodity and there is a dearth of RF spectrum, hence IoT connections are drifting towards Cognitive Radio Networks (CRNs). Permeating things with cognitive abilities will be able to make RF spectrum decisions to achieve interference-free and wireless connectivity as per their QoS requirements. The wireless systems are rapidly advancing. The leap from packet switching along with circuit switching with 144 kbps data rate (2G and 2.5G) to Long Term Evolution Advanced (LTE-A), i.e., 4G occurred in one decade time frame. As the current wireless connectivity is aimed at higher capacity, higher data rate, low end-to-end latency, massive device connectivity, reduced cost and consistent Quality of Experience (QoE) provision, therefore, 4G is being replaced with 5G. Presently the Radio Frequency (RF) spectrum band is fully sold out and allocated to various wireless operators and applications. On the other hand, new wireless applications are emerging and there is a serious dearth of frequency spectrum to be allocated to emerging wireless services. The efficient utilization of assigned RF spectrum which is otherwise underutilized due to the typical usage by the licensed users known as Primary Users (PUs) is the one of the best possible way to implement IoT in 5G. Thus the Spectrum Decision by unlicensed users of CR holds a significance in CR-based IoT in 5G and beyond network. This chapter describes a scientific supported spectrum decision support framework for CR Network. The main goal of this chapter is to discuss how CR technology can be helpful for the IoT paradigm.

Keywords: cognitive radio, internet of things, 5G/B5G, Spectrum decision framework, primary user, IoT-User

1. Introduction

4G provides voice, data and multimedia imparting to the wireless subscribers on every time and everywhere basis at higher data rates in Multimedia Messaging Service, Digital Video Chat Broadcasting (DVB), video chat, High Definitive TV content and mobile TV. As an application of 5G, the wireless systems are deployed to make All the people and things to be connected Any time with Anyone while being Anywhere via Any Path or Network and Any service (A6 connection). This A6 connection is known as Internet of Things (IoT). Internet of Things (IoT) is the environment where all over smart interconnected objects are connected with each other through unique addressing schemes based on specific telecommunication standards and protocols [1]. IoT based devices are to be interconnected through Base Transceiver Stations (BTSs) in wireless operations and the BTSs are linked with backhaul connectivity through Optical Fiber Transmission systems achieving higher bandwidths supplemented by Terrestrial Microwave links. The wireless Radio Frequency spectrum (WRFS) is almost completely assigned to existing wireless applications. At the same time, WRFS is underutilized due to the typical usage of mobile and other wireless services. To address this problem, Cognitive Radio (CR) has emerged as an enabling technology which offers a solution to spectrum scarcity problem. Hence, the CR-based IoT system has by default becomes a focus for researchers in wireless communication systems. CRN systems have emerged as a capable solution to the spectrum scarcity and as an enabling technology for the optimum utilization of otherwise underutilized RF spectrum, humanizing the synchronicity and interoperability in various wireless and mobile communications systems transforming into telecommunication devices and systems autonomous and self-reconfigurable. With swift shift to smart communication technologies and infrastructure, the Internet of Things (IoT) has emerged as a modern challenge in international Telecommunication industry and wireless applications. SUs access RF spectrum bands in heterogeneous manners in CRN and IoT supported smart area consists of heterogeneous devices, which are mobile as well as static in nature [1]. At the same time, the next generation mobile communication network referred to as the Fifth Generation (5G) is almost realized in the advanced telecommunication era [2]. 5G and beyond is expected to integrate the contemporary wireless technologies into an all Internet Protocol (IP) based networks which offer high performance worldwide network [3]. As the bandwidth for 5G and beyond is very large and the WRFS offers a large number of non-continuous idle spectrum slots in 5G communication as well [4], there is a requirement to identify the unused spectrum slots not being used by respective licensed users called primary users (PUs). This process is known as Spectrum Sensing (SS) in Cognitive Radio systems. Accurate SS allows the secondary users (SUs) to opportunistically use the vacant spectrum slots as per their wireless applications and vacate when the PU arrives in the network. This process is termed as spectrum decision. When optimally done, the SU along with PU will be enabled users using IoT paradigm in 5G/B5G networks. Therefore, spectrum decision is an important parameter for the deployment of CR-based IoT in 5G/B5G network.

All previously mentioned research contributions are more concept oriented for IoT in 5G network. The RF spectrum accessibility as per the wireless application for the user in IoT environment remains an open research area. Motivated by this, a comprehensive survey on 5G networks embedded with IoT applications based on CR ensuring A6 connectivity by accessing across the entire RF spectrum has been carried out in this chapter. A case study based on this survey for CR-based IoT in 5G networks has also been proposed to validate the concept.

2. Structure of the chapter

Introduction, related work and the motivation of the work is given in the first section. Evolution to 5G is given in Section 3. IoT in 5G network is described in Section 4. Section 5 gives an account for 5G with CRN based IoT and the need for RF spectrum management is given in Section 6. Section 7 concludes the chapter. **Table 1** given below lists the abbreviations used in this chapter.

Abbreviation	Definition	Abbreviation	Definition
AWGN	Additive White Gaussian Noise	MF	Membership function
AMPS	Advanced Mobile Phone Service	MIMO	O Multiple input multiple output
B3G, B4G, B5G	Beyond third, fourth and fifth generations	MISO	Multiple input single output
BTS	Base Transceiver Station	MHz	Mega hertz
CCC	Common control channel	MVR	Majority vote rule
CDMA	Code Division Multiplexing Access	NB Narrow band	Narrow band
CR	Cognitive Radio	NI	National Instrument
CRN	Cognitive Radio Networks	OFDM	Orthogonal frequency division multiplexing
DSA	Dynamic Spectrum Access	PC	Personal Computer
		PDP	Poisson distribution process
DSMF	Dynamic Spectrum Management Framework	PSD	Power spectral density
DVB	Digital Video Chat Broadcasting	PU	Primary user
D2D	Device to Device	QoE	Quality of experience
IoT	Internet of Things	QoS	Quality of service
ED	Energy Detection	QPSK	Quadrature phase shift keying
EV-DO	Evolution Data Only/Evolution Data Optimized	RASC	Random channel assignment with single channel
ETACS	European Total Access Communication System	RFID	Radio frequency identification
FSDM	Frequency Spectrum Decision Mechanism	ROC	Receiver operating characteristics
HART	Highway Addressable Remote Transducer Protocol	SDR	Software define radio
RFID	Radio frequency identification	SDSF	Spectrum decision support framework

Abbreviation	Definition	Abbreviation	Definition
ROC	Receiver operating characteristics	SG	Smart grid
6TiSCH	IP (IPv6 settings) integrated with Time synchronized channel hopping		
SNR	Signal to noise ratio		
SU	Secondary user		
TDMA	Time division multiple access		
UE	User equipment		
URLLC	Ultra-reliable low latency communication		
UWB	Ultra-wideband		
WCDMA	Wide-band code division multiple access		
WiFi	Wider Fidelity		
WiMAX	Worldwide interoperability for microwave access		
WLAN	Wireless local area networks		
WWRF	Wireless world research forum		

Table 1. Abbreviations used.

3. Advancement of wireless technologies from 1G to 5G

Since the inception of the first generation (1G) cellular systems in telecommunication system, the entire pattern of living environment including the people's work, lifestyle and the agricultural and industrial development trends has been effected. Evolution to the fifth generation (5G) is the progressive advancement in the telecommunication industry to keep with the growing pace of mobile data traffic, huge volume of device connections and continuous emergence of latest commercial scenarios. Over the last one decade or so, the wireless communications have the capability to connect all the existing mobile technologies, to build a terminal that is to support the voice, video and data applications with respective QoS requirements guaranteed i.e., at very high data rates and users speeds making it a 5G/B5G environment. The chronological evolution to 5G/B5G [5] is listed here in the **Table 2**.

Device to device (D2D) communications in 3GPP and LTE standards offers transfer of data directly to each other without the involvement of BSs [6]. This reduces the workload and energy consumption of BS thereby offering a good platform for 5G. Emerging 5G wireless communications envision very high data rates (typically of Gbps order), extremely low latency, significant increase in BTS capacity and improvement in PUs' and SUs' perceived Quality of Experience (QoE), compared to existing 4G/3G wireless networks. The 5G/B5G implies the whole wireless world interconnection (WISDOM; Wireless Innovative System for Dynamic

Wireless technology generation	Applications	Standards	Data rates	Mobility offered	Time span
1G (Analog)	1st Generation of the mobile telecommunication technology standardized by the voice service.	NMT, AMPS, TACS, ETACS and JTACS	14.4 kbps	Low Speed	1995–1997
2G (Digital)	2nd Generation of wireless telephone technology introducing a data service; SMS (short message service)	TDMA, GSM, CDMA, 2.4 GHz narrowband WLAN	144 kbps	Low and medium speeds	1997–2000
3G (IMT 2000)	3rd Generation of mobile telecommunications (International Mobile Telecommunications-2000)	CDMA2000, EV-DO, W-CDMA, 802.11 PAN, Bluetooth.	384 kbps	Medium and High Speed	2000–2005
B3G	Beyond 3rd Generation	WiBro802.16e, WiMax, 3GPP, LTE	<50 Mbps	High Speed	2005–2010
4G	4Th Generation of mobile telecommunications	DAB/DVB, cellular GSM, IMT-2000, WLAN, IR, UWB, DSL, LTE-A, IEEE802.16e	<100 Mbps	Very High Speed	2010 onwards
5G/B5G	5th Generation and beyond.	4G + WISDOM			2015 onwards

Table 2. Progressive evolution of Mobile services from 1G to 5G [5].

Operation Megacommunications concept), with guaranteed QoS requirements of wireless services [7]. Spectrum decision in CR would ensure spectrum scarcity problems and IoT complies wireless A6 connections for users, making CR-based IoT in 5G networks with a focus on spectrum decision framework in CR for IoT in 5G networks, an interesting study for researchers.

4. IoT in 5G/B5G networks

Mobile data and IoT are the future internet for everything and will be the key and motivating force in the advancement of 5G/B5G networks. In the time to come, likely by 2021–2025, 5G/B5G will not only meet the assorted requirements of people in various constituencies of daily life such as residence, work, leisure, and transportation, but also will infuse the IoT and light up the diverse specialized domains to the professional aspects of human life and the industry such as medical sciences and facilities and transportation to realize the true interconnectedness of all things [9]. The realization of IoT is dependent on internet application scenario based requirements which converge to 5G networks and are not guaranteed in 4G and LTE technologies. These requirements are listed in **Table 3**.

Internet application situation	Mobile data provided internet to the subscribers		IoT	
	Wide and seamless coverage	**High capacity to guarantee QoS requirements for Internet Applications**	**A6 connection**	**Low end to end latency**
Requirements	Seamless connectivity with high speed service in mobility of the subscriber	Enormously high data transmission rate	Provision connectivity to billions of devices with matching capability of power requirements for devices	Provision of service to users with less than millisecond end-to-end delays in transmission and in switching of spectrum slots.

Table 3. Modern trends and requirements in IoT [8].

5. 5G/B5G with CRN based IoT

The Internet of Things (IoT) envisions thousands of constrained devices with sensing, actuating, processing, and communication capabilities able to observe the world with an unprecedented resolution. According to Cisco, more than 50 billion devices are expected to be connected to the internet by 2020 and 20% of which are from the industry sector [9]. These connected things will generate huge volume of data that need to be analyzed to gain insight behind this big IoT data. Moreover, in the industrial environments (industry 4.0) as well in in smart spaces (building, houses, etc.) and connected cars communications often require high reliability, low latency and scalability. Several technologies such as BLE, Zigbee, Wireless HART, 6TiSCH, LPWAN (Lora, Sigfox, etc.) have been proposed to fit these requirements. The forthcoming 5G networks are promising not only by increased data rates but also low-latency data communication for latency-critical IoT applications. 5G will enable massive IoT devices connected via a myriad of networks and critical machine type communications. While the massive IoT is more concerned about scalability deep coverage and energy efficiency, the later requires ultra-low latency and extreme reliability (URLLC). Recently, the fog-to-thing continuum [10] is proposed to mitigate the heavy burden on the network due to the centralized processing and storing of the massive IoT data. Fog-enabled IoT architectures ensure closer processing in proximity to the things, which results in small, deterministic latency that enables real time applications and enforced security. The IoT is a modern and the state of the art archetype in the technological advancement which is evolving as a future Internet. As per the principal vision of the IoT, the further requirement is the ubiquity of the Internet, after connecting people anytime and everywhere, is to connect extinct entities. By providing objects with embedded communication capabilities and a common addressing scheme, a distributed and permeating network of impeccably connected diverse electric and electronic devices is designed, which is to be indigenously cohesive into the existing Internet connections and mobile networks. Formally, IoT can be defined as, " A worldwide network on electronically interconnected devices uniquely addressable, based on standard communication protocols and allows users to be A6 connected" [11]. Thus allowing for the development of new intelligent services available anytime, anywhere, by anyone and anything. Latest research work and

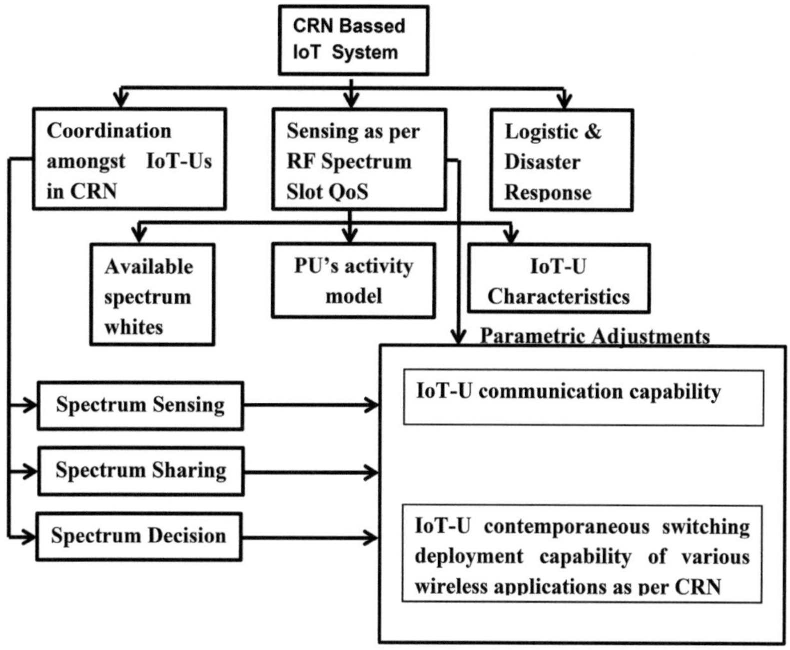

Figure 1. CRN system with its properties and research directions enabling it for IoT system for A6 connections by SUs.

technological systems are converging towards IoT and CRNs. Since the spectrum assignment policy involves expenditures for buying the RF spectrum, the assignment of spectrum for a huge number of devices and objects required for IoT connectivity will result in redundant cost effects. CRNs due to its typical spectrum utilization characteristic emerge in realization of IoT. The idea of a reserved spectrum slot as shared-to-reserve (SR) and reserved-to-share (RS) schemes in CR-HetNets proposed in [12] can enhance the system throughput and would offer a high bandwidth transmission for IoT-Us in CRN. A CRN properties enabling it for IoT applications is shown in **Figure 1**.

Usually, the SU operates in half-duplex mode (HD), i.e., it can either transmit or sense at any instant of time [13]. Due to this HD operation of SU, there is a possibility that harmful interference to PU is caused on unexpected arrival of PU and its activity during the transmission of SU. Hence the spectrum sensing should be a continuous process and SU must vacate the licensed channel on arrival of its PU and switch to another suitable channel as per its application, i.e., a befitting spectrum decision framework is essential.

6. Need for Spectrum management in IoT based 5G/B5G networks

External storage solutions offer nearly unlimited capacity, with dedicated signal processing to sort through data and find signals, interactions or events of interest. These long-duration, high-bandwidth solutions are ideal for today's crowded spectrum and advanced technologies such as cognitive radios. WRFS is characterized by PU activity modeling and accurate SS [14]. This means that the spectrum management holds a great significant in CR technologies and A6 connection. The DSA allows the users (both PUs and SUs) to optimally use the spectrum slots while guaranteeing their QoS requirements. Preserving the required QoS of the users along with their mobility requires spectrum mobility for the SUs in the network, which we now know as 5G network. Because of its mobility, an SU may change its location (cell) in a cellular network during its transmission and, therefore, will enter a new region in which the targeted RF spectrum slot is already being used by the PU [15]. The perfect SS techniques provide prior information for which SUs will work [5]. Since the primary traffic in any cell and region is always time varying and cannot be accurately predicted, therefore, the SUs must have the real time information of RF spectrum slots occupancy status to switch over to the vacant slots for resuming their transmission in case PUs arrive. Similarly, the SS errors are also required to be mitigated. The PUs use their licensed spectrum for their transmissions as per their QoS requirement and the statistical analysis says that this usage remains for a very short period of time. The decision of accessing the vacant spectrum slots would enable the SU to have A6 connections making an IoT environment. Therefore, SU is renamed here as IoT-User (IoT-U).

The allocated frequency spectrum for wireless applications is under-utilized due to the emblematic customs of wireless applications [16]. The conventional approach to spectrum management is very inflexible in the sense that each wireless service provider is assigned an exclusive license to operate in a certain frequency spectrum band. It has become very difficult to find vacant spectrum bands (to either deploy new services or to enhance existing ones) [17]. Therefore, for efficient utilization of the spectrum creating opportunities, Cognitive Radio (CR) technology allows its users called Secondary Users (SUs) to occupy the available spectrum slot for their communication (model is shown in **Figure 2**) and vacate it on arrival of the licensed user (called Primary User (PU)) [18]. The process of spectrum utilization using CR systems requires a Dynamic Spectrum Management Framework (DSMF) [19] which consists of four main components (naming Spectrum Sensing (SS), Spectrum Decision Framework, spectrum sharing and spectrum mobility). Spectrum sharing refers to coordinated access to the selected channel by the SUs. Spectrum mobility enables SU to switch over to another channel when a PU is arrived. SS involves identification of spectrum holes and the ability to quickly detect the onset of PU communications in the channels occupied by the SUs. Spectrum decision enables SUs in CR Network (CRN) to select the best available spectrum slots to satisfy SUs' Quality of Service (QoS) requirements without causing harmful interference to PUs [20]. On appearance of PU in the spectrum slot in which SU was carrying out its transmission, the SU looks for (which requires an efficient SS) and selects (which requires a suitable Spectrum Decision Framework) another QoS complying spectrum slot which is available/vacant. This implies that there are two steps (SS and spectrum decision framework) of efficient utilization of spectrum in CR-based IoT in 5G/B5G.

There are two scenarios as an overlay and underlay for spectrum sharing [21, 22]. In spectrum overlay scenario, a SU accesses a RF spectrum slot only when it is not being used by the PU [23]. This scenario is also known as opportunistic spectrum access. In other scenario the SU coexists with the PU and transmits with power constraints to guarantee the quality of service (QoS) of the PU. This scenario is known as underlay spectrum sharing [24]. The overlay mode operation is focused in this chapter.

http://dx.doi.org/10.5772/intechopen.80991

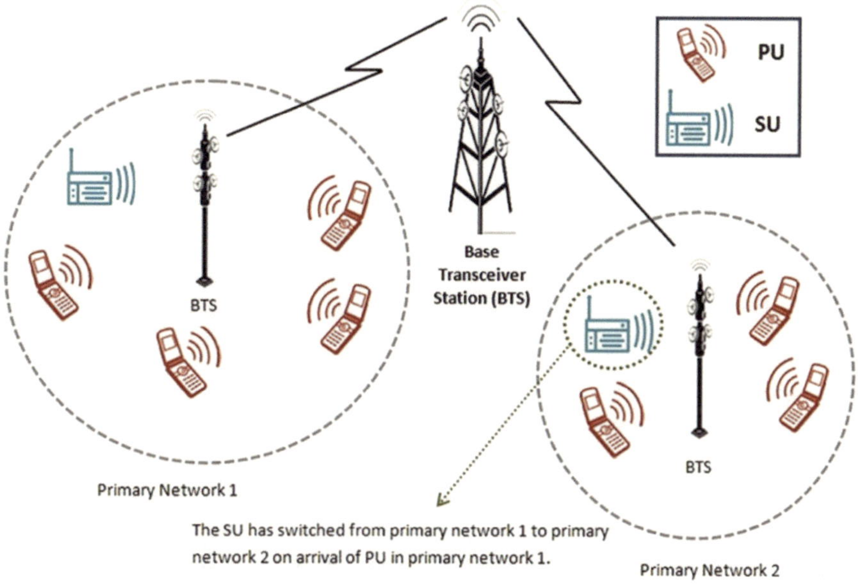

Figure 2. Working of a CRN model in which SU switched to another network on arrival of PU in primary network 1.

6.1. Spectrum sensing

CR systems offer the capability of IoT-Us to improve the spectrum utilization under the existing fixed spectrum assignment policy. IoT-Us cannot only sense the spectrum environment around and access vacant spectrum slots in the opportunity way, but also require to sense the presence of PU's signal continuously to keep the SS data updated. Hence, SS is the fundamental requirement in CR and the foundation for Spectrum Decision. The SS techniques are categorized into non cooperative (Energy Detection (ED), Matched Filter Detection (MFD) and cyclostationary detection), interference detection and cooperative (centralized SS (CSS) and non-centralized detection) [25, 26]. ED is the widely used scheme for SS due to its simplicity, easy implementation and it corresponds to the general purpose of SS for heterogeneous wireless communication systems [27]. The improved ED and CRs with multiple antennas can increase the SS performance [28]. MFD requires the exact synchronization and prior knowledge of PU signal, moreover implementation complexity of the sensing unit is large as the SU need receivers for all types of signal [29]. ED and MFD perform non coherent (by calculating the energy of the received signal samples) and coherent (comparing with the known PU signal) detection respectively [30]. Cyclostationary detection suffers from high complexity as all the cycle frequencies are required to be calculated [31]. CSS and non-centralized detection have exhibited SS errors due to time lag involved between sensing and its results [32]. Although the vacant spectrum slots are identified in SS but unless these are not simultaneously occupied through a well-defined decision process, the concept of CR cannot be realized. Therefore, it is imperative to mitigate the SS errors (false alarm and miss detection) before taking the decision to occupy the sensed vacant spectrum slot. A brief analysis of existing SS techniques [33] is given in **Table 4**.

6.2. PUs and IoT-Us activity model

In CRNs, there are two types of users to use the WRFS, one is PU and the other is SU, which we have renamed as IoT-U in this chapter. Since FCC has approved the access of unlicensed users (IoT-Us) to the already sold RF spectrum provided the unlicensed users do not cause harmful interference to PUs. The performance of CRNs is largely dependent on PU arrival and departure from the spectrum slots, the license of which it holds [34]. Hence, it is very important to model the PU activity for CRNs to enable IoT-Us to decide for occupation idle spectrum slots. PUs in the wide range (kHz to GHz as UWB and 5G networks operate in 3.6–39 GHz) of WRFS operate in any spectrum depending upon the specific wireless applications. **Table 5** shows the operating radio frequency bands for various wireless technologies.

As the RF spectrum band is wide range for various wireless applications, therefore, one PU activity cannot reflect the activity pattern of PU of all wireless applications as these varies from application to application. As the FCC has approved to use secondary users on licensed RF spectrum only with the condition that PU transmission will not be interfered. This implies that the licensed spectrum will only be occupied when PU is not using it, the underlay occupancy. Moreover, it is very important to ensure that PU is not harmfully interfered. That is why, the CRNs' performance is dependent on PU activity. Stochastic geometry provides a natural way of defining and computing macroscopic properties of mobile users' networks, by averaging over all potential geometrical patterns for the nodes, in the same way as queuing theory offers mobbing and the reliability, i.e., low end-to-end latency in wireless communication, average out the overall possible arrival patterns of the PUs within the networks on assigned RF spectrum. Thus PU activity modeling in wireless communication networks in terms of stochastic geometry is particularly relevant for spectrum decision framework. The PU activity, as a simplest case, in CRN can be represented as a print of a stationary random model in a probabilistic way. In particular the locations of the CRNs nodes are seen as the realizations of some point processes. When the underlying random model is ergodic, the probabilistic analysis also provides a way of estimating spatial averages which often capture the key dependencies of the CRN performance characteristics (connectivity, stability, capacity, etc.) as functions of a relatively small number of parameters [35]. Hence, the PU activity should be modeled with some stochastic arrival and departures probability expression. Poisson distribution process (PDP) provides a near to realistic probability of arrival and departure of the (primary) user in the network. PDP offers spatio-temporal representation of PU activity model. Moreover, Poisson distribution process is simple and adapts well in wireless communication scenario. The PU's arrival and departure follow the poison distribution process:

$$P(k\ events\ in\ interval) = \frac{\lambda_p^k e^{-\lambda_p}}{k!} \quad (1)$$

where k is occurrence of PU arriving and takes values *0,1,2,3, ...,* N and λ_p is the arrival rate of PU in the spectrum slot in the CRN. The existing spectrum decision techniques model PU activities without taking into account for SU (IoT-U) behavior and characterization. This is desirable in this CRN growth era as well as in CRN based IoT-U to have its model defined. This will help in ensuring no interference and will provide basics for mechanism of switching

SS Techniques	Method Used	Main Feature
ED, MFD and Cyclostationary detection based SS	Based on PU's transmitter	ED does not require the prior information of the PU
		MFD is related to prior knowledge of PU's signal
		Cyclostationary detection relies on distinguish between the PU's signal and the noise
Cooperative SS	Combining the sensing results of multiple SUs to improve the detection reliability	The fusion mechanisms including reliability based cooperative decision fusion. One is described in section of this chapter
		Using directional antenna
		Quashing interferences
		Integrates quickest detection and belief propagation framework
		Guaranteeing the high sensing accuracy in vehicular networks or industrial wireless networks
Spectrum-database SS	Enables to find all available spectrum slots by comparing the historical information of spectrum usage pattern with the received by a base station from each SU in the network	Exploiting spectrum table for SS
		A framework for determining the topology of vehicular network
		An iteratively developed history processing database
		A mobile crowd sensing-driven geolocation spectrum database for D2D communication
Compressive SS	Each SU detects and extracts the wide band signal directly to achieve efficient wide band sensing with much lower sampling rate than the Nyquist Criterion	Wideband SS scheme
		Based on real time PU's signal
		Analyze the sparsity of the wideband spectrum
		Reducing SS errors
		Spectrum occupancy status measurement
Full duplex SS	Each SU in the network can access the vacant spectrum slots while sensing the spectrum continuously	Listen and talk protocol
		Joint mode/rate adaptation policy for WiFi/LTE-U
		At low SNR values
		Optimal detection thresholds
		Canceling the self-interference of transceiver

Table 4. List of SS techniques available in literature [33].

Wireless applications	Frequency spectrum bands	Bandwidth
IEEE 802.11 g to n/WiFi	2.4 GHz	10 KHz
IEEE 802.16/LAN/2	5 GHz	100 KHz
IEEE 802.22	54–862 MHz	5–20 MHz
GSM	890–915 MHz (uplink) 935–960 MHz (downlink)	200 KHz
CDMA	800 and 1.9 GHz	125 MHz
W-CDMA	850–2100 MHz	125 MHz, 250 MHz
LTE	1710 –1770 MHz (uplink) 2110–2170 MHz (downlink)	20 MHz
UWB	3.1–10.6 GHz	500 MHz
5G Cellular	26.5–40 GHz and 30–50 GHz	All ranges of bandwidths i.e., narrow, wide, ultra wide and super ultra wide bands communication systems

Table 5. Frequency spectrum ranges for various wireless applications.

to other available slot, if PU arrives. The study of opportunistic spectrum access in CRNs with SU's transmission performance reveals that the interference caused to PU can be avoided by evaluating the SU's transmission blocking [34]. When PUs appear in the multiple spectrum slots in a WRFS band denoted by '*S*', IoT-Us need to vacate the spectrum slot and switch to another suitable spectrum slot (to complete their transmission) without interfering the PUs. This transmission process for PU and IoT-U is shown in **Figure 3**. This causes IoT-Us a temporary break in transmission, which is mitigated by simultaneous access in multiple noncontiguous spectrum bands by IoT-Us for their transmission. Even if a PU appears in one of the channels, the rest of the channels will continue to allow SUs to transmit while maintaining their QoS requirements.

In the transmission process of SUs when accessing the channel in a heterogeneous manner, the transmission level measure for PU is given by the PTB as,

$$P_{TB} = P(i,j) = \frac{\lambda_p (P_{i-i,j}) \varphi(i+1.j) + (j+1)\, \mu_s P(i,j+1) \varphi(i.j+1)}{i\,\mu_p + j\,\mu_{s+} \lambda_p + \lambda_s} \tag{2}$$

where PUs and IoT-Us arrive and depart from each spectrum slot in *s* at the rate λ_p and λ_s respectively. Similarly, the μ_p and μ_s are the mean values of the respective transmission durations of PU and SU in the network. The number of spectrum slots in *s* by PU and IoT_U at some specific time are represented by *i* and *j* respectively, such that $i + j \leq N$. P (i,j) is the stationary probability of two dimensional Markov state which is P_{TB}.

The state space ω for PUs and IoT-Us occupying spectrum slot in *S* is given as under;

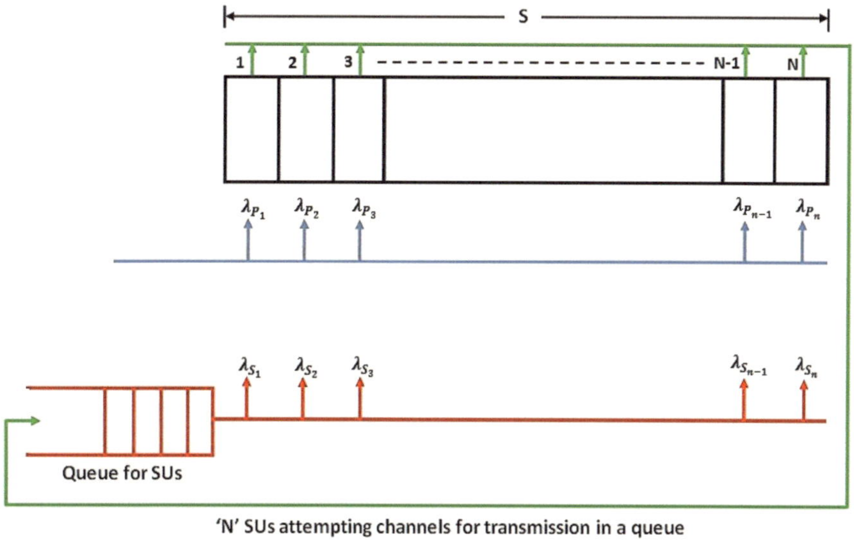

Figure 3. Transmission process for PUs and SUs (IoT-Us) to mitigate interference.

$$\bar{\omega} = \{(i, j) \mid 0 \leq i \leq N; 0 \leq j \leq N\} \tag{3}$$

and

$$\varphi(i, j) = \begin{cases} 1, (i, j) \in \bar{\omega} \\ 0, \text{otherwise} \end{cases} \tag{4}$$

7. Spectrum decision framework

CR Technology is characterize by PUs and SUs, a coordination among the IoT-Us to use the licensed bands when PUs are not using spectrum slots in the targeted spectrum slot. Although, in the recent 5 years, the researchers have carried out work in the area of Spectrum Decision, however, still this area is not yet fully explored. A survey of spectrum decision in CRNs based on RF spectrum characterization, spectrum selection and CR reconfiguration has been presented in [20]. Since then (2013) in one of spectrum decision research works, the authors have proposed a fuzzy inference-based decision strategy which is based on three key parameters, spectrum slot idle time, spectrum slot occupancy status and spectrum slot QoS [36] which provides an accurate and robust spectrum decision framework for SUs. The same and can be equally effective for IoT-Us in CR-based IOT in 5G/B5G networks as it encompasses all signal processing matrices such as Massive MIMO Antennas, varying bandwidths characteristics and wireless propagation channel models for all wireless applications (through spectrum slot QoS), required for wireless communications. Likewise, a spectrum decision scheme is

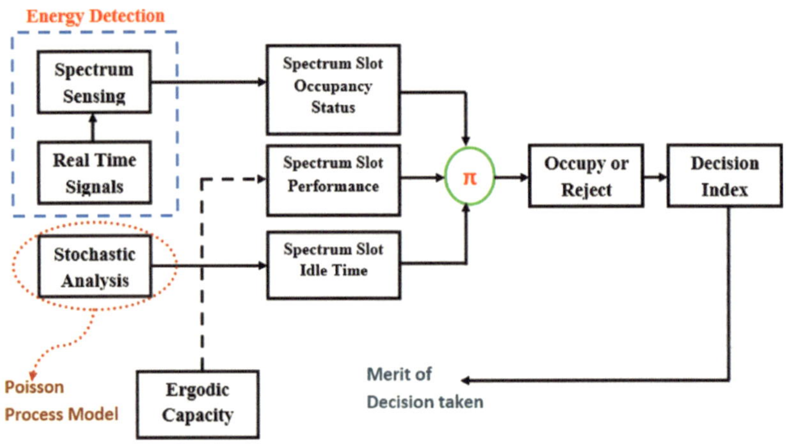

Figure 4. Proposed Spectrum decision framework for CRN based IoT in 5G/B5G networks.

proposed here with an analysis through Radio Operating characteristics (ROC) curves at various SNR values. This scheme is based on fusion of three separate decision of three key parameters, spectrum slot idle time, spectrum slot occupancy status and the spectrum slot performance. The spectrum decision framework first finds the idle time of the spectrum slot, spectrum slot occupancy status through SS using ED scheme and spectrum lot performance based on its ergodic capacity as shown in **Figure 4.**

7.1. System model

A CRN operating in a spectrum band '*S*' with frequency ranging from 54 MHz–50 GHz covering most of the wireless applications given in **Table 5**, is considered. The other specifications to be considered (like uplink and downlink frequencies, modulation techniques used in transmission and bandwidth) are as given in [37]. Channel bandwidth is the frequency range over which a IoT-U's transceiver transmits and receives its signals in CRN. An IoT-U can carry out its transmission on either narrow, wide and ultra wide band (UWB) ranges depending on the RF environment and wireless applications. The CRN has a centralized network operator, for instance a BTS which functions as "serve to provide". A region comprising of 5 BTSs, unlimited number of mobile devices, all buildings in the neighborhood are under the coverage of all the wireless services as shown in **Figure 5**. A wide range wireless based applications, i.e., GSM, bluetooth, UWB, NB, video conferencing, IP based communication, office automation systems, building security management systems, 5G and RFID, connected through IP based communication radios. 3GPP channel model has been used owing to its typical characteristics for wireless systems, i.e., it has properties that impact on system performance by reflecting the important properties of propagation channels. Moreover, wireless networks are optimized in the region of system model. Let there be '*J*' SUs (using ED for SS) in the CRN each having its own Software Defined Radio (SDR) to exploit the multiple spectrum bands over wide spectrum ranges by adjusting the operating frequency through software operations. The BTSs exercises control over all *J* IoT-Us within its transmission range as shown in **Figure 2**. The

http://dx.doi.org/10.5772/intechopen.80991

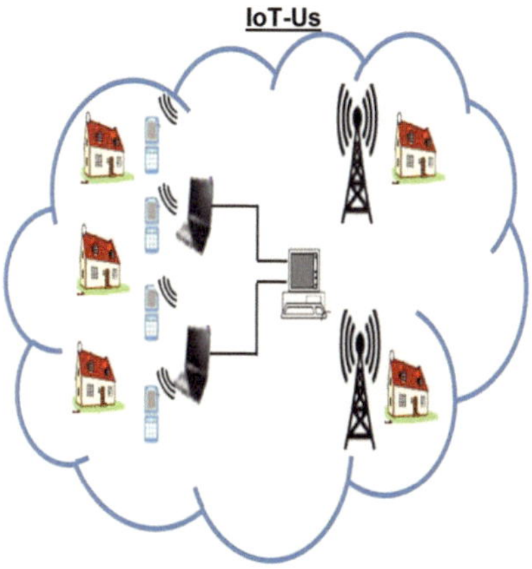

PUs' Licensed Access to Wireless Applications

Services	RF Spectrum Band		Bandwidth
GSM	Uplink 890 MHz	Down Link 95 MHz	200KHz
TVWS, IEEE 802.22	54 MHz-862 MHz		6,7 & 8 MHz
Zig Bee	850 MHz-930 MHz		Narrowband
Wi-Fi, IEEE 802.11a/g	2.4 GHz		
UWB	3.1 GHz-10.6GHz		500 MHz
PCS	Uplink 1710 -1780 MHz	Down link 1805-1860 MHz	30 MHz
US Paging	931 MHz-932 MHz		
GPS	L1:1375.42MHz L2:1227.60MHz		1MHz
ISM	902-928 MHz 2.4-2.484 GHz 5.725-5-850GHz		Narrowband
CDMA	Uplink 824-849 MHz	Down Link 869-894GHz	1.25 MHz
WCDMA	1900-1920 MHz	2010-2025 MHz	5 MHz

IoT-U
Configures its radio as per Characteristics of
Wireless applications for A6 connections

Figure 5. System model for proposed Spectrum decision framework.

RF spectrum slots in CRN are considered to have varying bandwidth. There are *N* PUs having rights to access same number of corresponding spectrum slots. *J* IoT-Us are attempting to access these slots for their transmission. Wireless services employ a combined FDMA/TDMA approach for air interface.

7.2. Performance evaluation and numerical analysis

The simulation parameters are given in **Table 6**. In finding the idle spectrum slot through the PU activity time, there can occur two types of errors. One is, false alarm and the other is the miss detection. Later is due to sensing an idle spectrum slot as occupied, and the following is due to assuming an occupied slot as idle. The performance of the proposed spectrum decision framework is assessed here through ROC curves between probability of false alarm and miss detection. These two are the inter-related parameters in the proposed decision process. To

No. of BTSs	5
No of IoT-Us in CRN	*J*
No of PUs	5
RF spectrum range	890–915 MHz (GSM Band)
Wireless channel model	3 GPP
Bandwidth	200 KHz

Table 6. Simulation parameters.

have accurate data of PU(s) activity time and occupancy status, there should be low values of both the probabilities, which cannot be achieved as both are inter related to each other. Therefore, an optimal set of range must be obtained. Transmission at different values of SNR give different ROC curves for the PU's activity time and its occupancy in the spectrum slot. The lower the probability of miss detection (or higher the probability of detection) for a given probability of false alarm, the more reliable and accurate the detection would be, which is desirable for taking the decision by the SU(s) to occupy that particular spectrum slot in CRN.

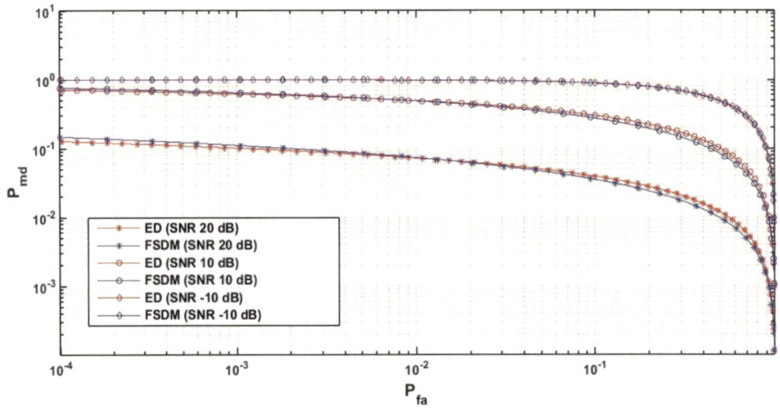

Figure 6. ROC curves for various values of SNR compared with ED SS results and on occupying the spectrum slots by 5 IoT-Us under proposed frequency Spectrum decision mechanism (FSDM).

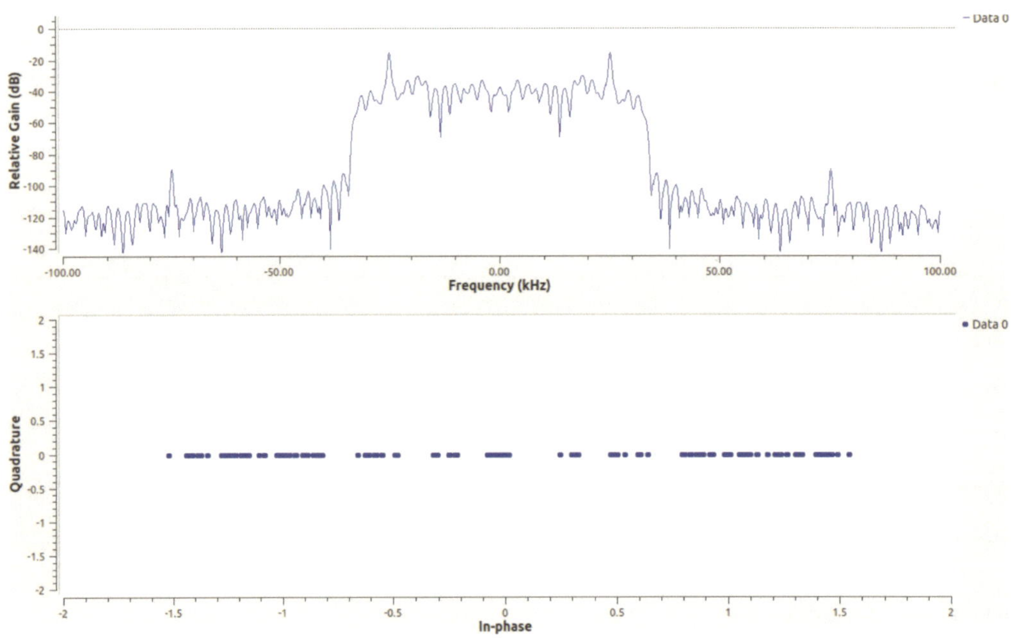

Figure 7. SU's transmitter transmitting low power signal around center frequency in their side lobs.

Energy Detection (ED) in SS offers a fast and reliable detection method for SUs in CRN. The detection performance of ED depends on effects of multipath fading [38]. Accordingly, the proposed decision framework has been validated by ROC curves compared with those of SS through ED method at various SNR values. ROC curves for 5 IoT-Us are shown in **Figure 6**. For the communication overhead as an outcome of information exchange required by the statistical approach, has been significantly reduced by using existing common control channels (CCC) by IoT-Us. This complexity cost is fully justified given the significant performance improvement that the proposed framework offers in terms of latency, throughput, energy efficiency, delay, and the reliability for realization of IoT in terms of A6 connections.

To ensure there is no harmful interference caused to PU by the SU, the SU's transmitter transmits less energy in the side lobs of the transmission signal as shown in **Figure 7**.

8. Conclusion

CR is an important measure to spectrum scarcity problem. To optimally utilize the already allocated spectrum, spectrum decision holds significance in CR. Spectrum decision enables CR users to access the spectrum slots as per their wireless application over a wide RF spectrum range. In this chapter a spectrum decision framework has been proposed which weighs the spectrum band on its idle time, occupancy status and performance and ensures A6 connection thereby providing an enable technology for IoT to support 5G/B5G networks with higher data rates.

Author details

Ahmad Naeem Akhtar*, Fahim Arif and Adil Masood Siddique

*Address all correspondence to: ahmadnaeem.akhtar@mcs.edu

National University of Science and Technology (NUST), Islamabad, Pakistan

References

[1] Atzori L, Iera A, Marobito G. The internet of things: A survey. Computer Networks. 2010:2787-2805

[2] Roy A, Sengupta S, Wong K-K, Raychoudhury V, Govindan K, Singh S. 5G wireless with cognitive radio and massive IoT. IETE Technical Review. 2017;**34**(suppl 1):1-3. DOI: 10.1080/02564602.2017.1414387

[3] Liu X, He D, Jia M. 5G-based wideband cognitive radio system design with cooperative spectrum sensing. Physical Communication. 2017;**25**(2):539-545, . ISSN 1874-4907. DOI: 10.1016/j.phycom.2017.09.010. (http://www.sciencedirect.com/science/article/pii/S1874490717302483)

[4] Shree KS et al. Dynamic spectrum sharing in 5G wireless networks with full-duplex technology: Recent advances and research challenges. IEEE Communication Surveys and Tutorials. 2018;**20**(1, First quarter). DOI: 10.1109/COMST.2017.2773628

[5] Badoi CI, Prasad N, Croitoru V, et al. 5G based on cognitive radio. Wireless Personal Communications. 2011;**57**:441. DOI: 10.1007/s11277-010-0082-9

[6] Ghavami H, Shirivani Moghaddam S. Analysis of outage probability for In-band device to device communications underlying cellular network. International Journal of Communication Systems (IJCS), Wiley (ISI-JCR). 2017;**30**(14):1-8

[7] Agiwal M, Roy A, Saxena N. Next generation 5G wireless networks: A comprehensive survey. IEEE Communication Surveys and Tutorials. 2016;**18**(3):1617-1655, 3rd quarter

[8] Perera C, Zaslavsky A, Christen P, Geogakopoulos D. Context aware computing for internet of things: A survey. IEEE Communication Surveys and Tutorials. 2014;**16**(1, First Quarter)

[9] Cisco Visual Networking Index: Global Mobile Data Traffic Forecast Update, 2012-2017. https://newsroom.cisco.com/press-release-content?type=webcontent&articleId=1912960. [Accessed on 18 Jun 2018]

[10] Saha B. Fog computing – A Bridgw in cloud-to-thing continuum of IoT. Mobiloitte. 2018. Available online https://www.mobiloitte.com/blog/fog-computing-a-bridge-in-cloud-to-thing-continuum-of-iot/ Accessed on 3 Aug 2018

[11] Khan AA, Rehmani MH, Rachedi A. When cognitive radio meets the internet of things?. 2016 International Wireless Communications and Mobile Computing Conference (IWCMC), Paphos, 2016, pp. 469-474. DOI: 10.1109/IWCMC.2016.7577103

[12] Habibzadeh A, Shirvani Moghaddam S, Razavizadeh SM, Shirvanimoghaddam M. Modeling and analysis of traffic-aware Spectrum handover schemes in cognitive Het nets. Transactions on Emerging Telecommunications Technologies (ETT). 2017;**28**(12)

[13] Sabat S, Sharma PK, Gandhi A. Full-duplex cooperative Spectrum sensing with primary user activity in cognitive radio networks. IETE Technical Review. 2017;**34**(sup 1):4-14. DOI: 10.1080/02564602.2017.1396937

[14] Lee W-Y, Akyildiz I. A spectrum decision framework for cognitive radio networks. IEEE Transactions on Mobile Computing. 2011;**10**(2):161-174

[15] El Tanab M, Hamouda W. Resource allocation for underlay cognitive radio networks: A survey. IEEE Communication Surveys and Tutorials. 2017;**19**(2):1249-1276, Second-quarter. DOI: 10.1109/COMST.2016.2631079

[16] Akyildiz IF, Lee W-Y, Vuran MC, Mohanty S. A survey on spectrum management in cognitive radio networks. CSE Journal Articles. 2008:87

[17] Lee W-Y, Akyildiz IF. Spectrum-aware mobility management in cognitive radio cellular networks. IEEE Transactions in Mobile Computing. 2012;**11**(4):529-542

[18] Bohning D. Notice of Proposed Rule Making and Order. FCC, ET Docket no. 03-222; 2003. pp. 197-200

[19] Akhtar AN, Rashdi A, Arif F. Fusion based spectrum decision framework for cognitive radio users. 2015 IEEE 16th International Symposium on A World of Wireless, Mobile and Multimedia Networks (WoWMoM), Boston, MA, 2015. pp. 1-6. DOI: 10.1109/WoWMoM.2015.7158193

[20] Masonta MT, Mzyece M, Ntlatlapa N. Spectrum decision in cognitive radio networks: A survey. IEEE Communication Surveys and Tutorials. **19**(3):1924-1944

[21] Ghavami H, Shirvani Moghaddam S. Outage probability of device to device communications underlaying cellular network in Suzuki Fading Channel. IEEE Communications Letters. 2017;**21**(5):1203-1206

[22] Shirvani Moghaddam S. A novel resource reuse in device to device communications. Internet Technologies and Society (ITS2017) Conference, Proceedings of IADIS digital library, Sydney, Australia; 11-13 December 2017. pp. 35-42

[23] Zou J, Xiong H, Wang D, Chen CW. Optimal power allocation for hybrid overlay/underlay spectrum sharing in multiband cognitive radio networks. IEEE Transactions on Vehicular Technology. 2013;**62**(4):1827-1837. DOI: 10.1109/TVT.2012.2235152

[24] Le L, Hossain E. Resource allocation for spectrum underlay in cognitive radio networks. IEEE Transactions on Wireless Communications. Dec. 2008;**7**(12):5306-5315

[25] Zayen B, Hayar A, Nussbaum D. Blind spectrum sensing for cognitive radio based on model selection. In: Proceedings of 3rd International Conference on Cognitive Radio Oriented Wireless Networks and Communications; May 2008

[26] Haykin S. Cognitive radio: Brain-empowered wireless communications. IEEE Journal on Selected Areas in Communications. 2005;**23**(2):201-220

[27] Akyildiz IF, Lee W-Y, Vuran MC, Mohanty S. NeXt generation/dynamic spectrum access/cognitive radio wireless networks: A survey. Elsevier Computer Networks. 2006; **50**(13):2127-2159

[28] Lu L, Zhou XW, Onunkwo U, Li GY. Ten years of cognitive radio technology. EUROSIP International Journal of Wireless Communication Networking. 2012;**28**:1-16

[29] Axell E, Leus G, Larson EG, Poor HV. Spectrum sensing for cognitive radio: State-of-art and recent advances. IEEE Signal Processing Magazine. May 2012;**29**(3):101-116

[30] Shirvani Moghaddam S, Kamarzarin M. A comparative study on the two popular cognitive radio Spectrum sensing methods: Matched filter versus energy detector. American Journal of Mobile Systems, Applications, and Services. Oct. 2015;**1**(2):132-139

[31] Chin WL, Li JM, Chen HH. Low-complexity energy detection for spectrum sensing with random arrivals of primary users. IEEE Transactions on Vehicular Technology. Feb. 2016; **65**(2):947-952. DOI: 10.1109/TVT.2015.2405558

[32] Singh A, Bhatnagar MR, Mallik RK. Performance of an improved energy Detector in multihop cognitive radio networks. In: IEEE Transactions on Vehicular Technology. 2016;**65**(2):732-743. DOI: 10.1109/TVT.2015.2401332

[33] Hu F, Chen B, Zhu K. Full Spectrum sharing in cognitive radio networks toward 5G: A survey. IEEE Access. 2018;**6**:15754-15776. DOI: 10.1109/ACCESS.2018.2802450

[34] Sun H, Nallanathan A, Wang C-X, Chen Y. Wideband spectrum sensing for cognitive radio networks: A survey. IEEE Wireless Communications. Apr 2013;**20**(2):74-81

[35] Baccelli F, Blaszczyszyn B. Stochastic geometry and wireless networks, Volume I-Theory. In: Baccelli F, Blaszczyszyn B, editors. Foundations and Trends in Networking. Vol. 1. NoW Publishers; 2009. p. 150. Vol. 3: No 3-4, p. 249-449. 978-1-60198-264-3, 978-1-60198-265-0

[36] Akhtar AN, Ghafoor A, Rashdi A, Arif F. Fuzzification supported spectrum decision framework for cognitive radio networks. Transactions on Emerging Telecommunications. 2017;**28**:e3229. DOI: 10.1002/ett.3229

[37] Rappaport, Theodore S. Wireless communications: principles and practice (2nd ed). Prentice Hall PTR, Upper Saddle River, N.J.; Great Britain: 2002

[38] Dighem FF, Alouni M-S, Simon MK. On energy detection of unknown signals over fading channels. IEEE Transactions on Communications. 2007;**55**(1):21-24

Section 3

Emulation Attack and Interference Alignment in Cognitive Radio Systems

Chapter 5

Spectrum Sensing and Mitigation of Primary User Emulation Attack in Cognitive Radio

Avila Jayapalan and Thenmozhi Karuppasamy

Additional information is available at the end of the chapter

http://dx.doi.org/10.5772/intechopen.80328

Abstract

The overwhelming growth of wireless communication has led to spectrum shortage issues. In recent days, cognitive radio (CR) has risen as a complete solution for the issue. It is an artificial intelligence-based radio which is capable of finding the free spectrum and utilises it by adapting itself to the environment. Hence, searching of the free spectrum becomes the key task of the cognitive radio termed as spectrum sensing. Some malicious users disrupt the decision-making ability of the cognitive radio. Proper selection of the spectrum scheme and decision-making capability of the cognitive reduces the chance of colliding with the primary user. This chapter discusses the suitable spectrum sensing scheme for low noise environment and a trilayered solution to mitigate the primary user emulation attack (PUEA) in the physical layer of the cognitive radio. The tag is generated in three ways. Sequences were generated using DNA and chaotic algorithm. These sequences are then used as the initial seed value for the generation of gold codes. The output of the generator is considered as the authentication tag. This tag is used to identify the malicious user, thereby PUEA is mitigated. Threat-free environment enables the cognitive radio to come up with a precise decision about the spectrum holes.

Keywords: cognitive radio, spectrum sensing, PUEA, collaborator node, authentication tag

1. Overview

The introduction of wireless technique has led to the achievement of mobility and global connectivity through its advantages in flexibility, cost and convenience. Due to its rapid growth, there arises a demand for the spectrum. But analysis shows that there are portions

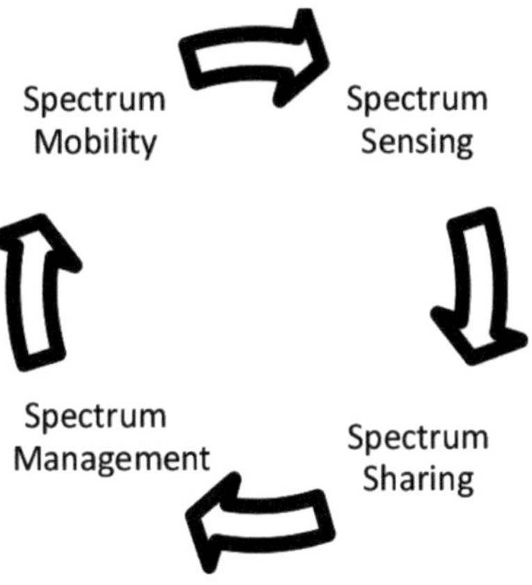

Figure 1. Cognitive cycle.

of the spectrum which are not effectively used and those portions of the spectrum could be exploited, whenever in need. For dynamic spectrum access, cognitive radio has risen as a favourable solution [1, 2]. Cognitive radio searches out for the free spectrum termed as 'spectrum holes'. The process of finding the spectrum holes is termed as spectrum sensing. Apart from spectrum sensing some of the other functions of cognitive radio are spectrum sharing, spectrum management and spectrum mobility. These four functions are put together termed as cognition cycle [3–6] and it is shown in **Figure 1**.

1.1. Spectrum sensing

The users in the wireless environment can be classified into three main groups, namely primary users, secondary users and selfish, malicious users. Primary users are those who gain ownership of the spectrum [7]. Secondary users desire to gain access in the absence of primary users [8]. Malicious users desire to own access of the spectrum by cheating the secondary users [9].

In the cognitive environment, the procedure of searching the spectrum holes by the secondary users is known as spectrum sensing. The cognitive radio not only looks for the free spectrum, but also checks for the arrival of primary users. On the homecoming of the primary users, cognitive users or the secondary users should quit the existing spectrum immediately and search for some other new spectrum hole.

1.2. Types

Various types of spectrum sensing schemes are available and they are shown in **Figure 2**. Some of them are energy detection method [10], cyclostationary method [11], matched filter method [12], etc. Feature detection and matched filter methods require prior knowledge about the licenced user for detection and they are time-consuming.

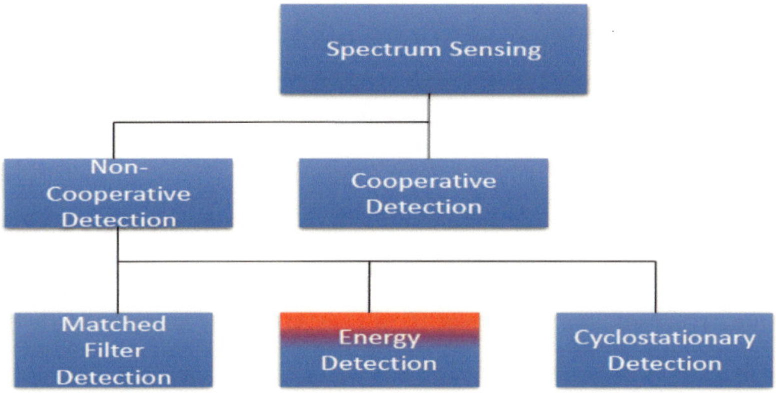

Figure 2. Types of spectrum sensing.

Energy detection method does not require any former knowledge about the primary user and it is simpler and quicker when compared to the previously mentioned methods. Energy detector can be classified into two types:

- Frequency domain-based energy detector
- Time domain-based energy detector

Energy detection method is not suited for places where the SNR is very low. Hence, it is a trade-off in choice of the proper spectrum sensing scheme.

1.2.1. Time domain

Figure 3 shows time domain-based energy detector. The energy of the signal is calculated and compared with the threshold.

The output of the detector is

$$Z = \sum_{n=0}^{N} y(n)^2 \tag{1}$$

where n = 1, 2, 3, …, N. N = number of samples

$$\begin{aligned} &\text{If } Z \prec \lambda \text{ primary user absent} \\ &\text{If } Z \geq \lambda \text{ Primary user present} \end{aligned} \tag{2}$$

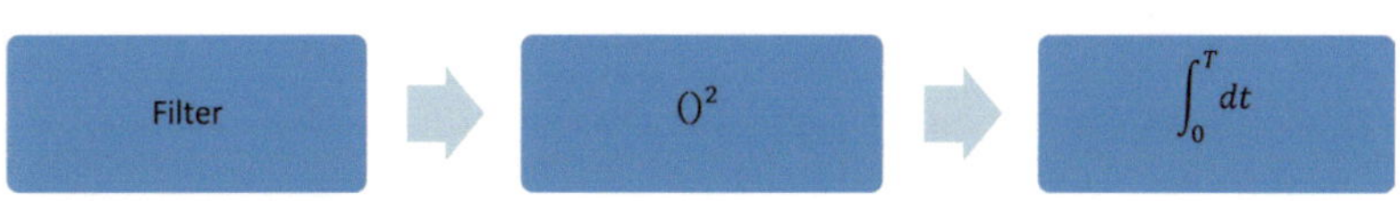

Figure 3. Time domain-based energy detector.

The decision hypothesis is as follows:

$$\begin{aligned} y(n) &= s\,H_0 = \text{only the presence of noise} \\ y(n) &= x(n) + s\,H_1 = \text{presence of both primary user signal and noise} \end{aligned} \tag{3}$$

where n is the noise, y(n) is the received signal and x(n) is the transmitted signal.

1.2.2. Threshold

Keeping the probability of false alarm fixed the threshold value is set according to the equation:

$$\lambda_f = \sigma_n^2\left(1 + \frac{Q^{-1}\left(P_f\right)}{\sqrt{\frac{N}{2}}}\right) \tag{4}$$

$$\lambda_d = \sigma_n^2(1 + SNR)\left(1 + \frac{Q^{-1}(P_f)}{\sqrt{\frac{N}{2}}}\right) \tag{5}$$

where N = number of samples and Q^{-1} = complementary error function.

- Cooperative spectrum sensing: Group of cognitive radios, shares the spectrum sensing information. To achieve spectrum sharing and to overcome the multipath propagation effects and hidden node problems cooperative spectrum sensing scheme is utilised. The cognitive users employ less sensitive detectors, thereby reducing the cost of hardware and complexity. It is divided into two types namely
- Centralised spectrum sensing
- Distributed spectrum sensing

Centralised spectrum sensing: In this method, the central unit collects the sensing information from the cognitive users located at various places of the radio environment, analyses the received information and transmits the final decision about the existence or nonexistence of the PU to the cognitive users. Two rules are followed in deciding PU. One is AND rule and the other is OR rule.

- AND rule: All the SU's declare that the PU is present
- OR rule: If anyone SU status is high then the PU is considered present

Distributed spectrum sensing: Each node senses the PU, and a decision is made based on the earlier scenarios. Complexity is greatly reduced as there is no need of fusion center (FC). But at the same time, it increases the burden to the CR.

1.3. PUEA

On receiving the primary users signal, the cognitive radio compares it with a predefined threshold. If the incoming signal exceeds the primary threshold, user is assumed to be present else absent. In the absence of the primary user, the malicious user sent a fake signal almost matching with the primary user signal to the cognitive radio. The cognitive radio on receiving the fake signal compares it with the threshold. The fake signal exceeds the threshold, and hence the primary user makes a wrong interpretation that the primary user is present and does not make any attempt access the spectrum. The malicious user now utilises that free spectrum. This attack is known as primary user emulation attack (PUEA) [13], which is considered as the severe attack in the physical layer of the cognitive radio.

Various researchers have analysed the importance and impact of PUEA in cognitive radio environment, and they have come out with different solutions to overrule PUEA. Few of them are as follows. A review about primary user emulation attack has been made in [14–17]. A study about PUEA has been made in [18, 19]. To ensure end-to-end security for portable devices over cognitive radio network, two authentication protocols have been proposed in [20]. Four dimensions continuous Markov chain model to combat PUEA has been proposed in [21]. PU, secondary user, selfish misbehaviour secondary user and misbehaviour secondary user are considered to combat PUEA. In [22], a trustworthy node is taken as reference and the position of PU and emulator was found to detect PUEA. Eigenvalue-based PUEA mitigating method has been discussed in [23]. Time-synched link signature scheme to mitigate PUEA has been proposed in [24]. In [25], temporal link signature scheme to establish link between transmitter and receiver has been proposed and with the aid of signature PUEA is mitigated. Any change in the transmitter location or emulator claiming as transmitter is identified.

Integrated cryptographic and link signature-based method to mitigate PUEA has been proposed in [26]. Suspicious level and trust level calculations are carried out to mitigate PUEA in cooperative spectrum sensing environment in [27]. Mitigating PUEA and worm hold attack through sequence number generation by the helper nodes has been proposed in [28]. Multiple helper nodes-based authentication method to combat PUEA in the TV band has been discussed in [29]. Optimum voting rule and sample-based scheme in cooperative spectrum sensing to mitigate PUEA has been proposed in [30]. Advanced encryption standard (AES)-based authentication method with 256-bit key size has been suggested in [31] to overcome PUEA. Digital constellation-based authentication scheme to mitigate PUEA has been proposed in [32]. Quadrature phase shift keying was considered. Based on the tag value, the phase of QPSK modulation is rotated. Helper node-based special authentication algorithm has been suggested in [33] to mitigate PUEA in mobile networks. Location, privacy-preserving framework, has been proposed in [34]. The framework consists of two parts namely privacy-preserving sensing report aggregation protocol and distributed dummy report injection protocol.

Authentication scheme based on the transmitter called localisation based defence (LocDef) to mitigate PUEA has been discussed in [35]. In [36], neural network and database management-based scheme to mitigate PUE threat have been proposed. COOPON (called cooperative neighbouring cognitive radio nodes) technique to mitigate the selfish user attack in cooperative spectrum sensing environment has been proposed in [37, 38]. Matched filter-based spectrum sensing together with the cryptographic signature-based method has been suggested in [39]. Extensible authentication protocol and carousel rotating protocol-based authentication scheme have been proposed in [40]. Location-based authentication protocol for IEEE 802.22 wireless regional area network (WRAN) has been implemented in [41]. Double key-based encryption scheme has been proposed in [42] to overcome the attacks. Two non-parametric algorithms namely cumulative sum and data clustering-based method have been discussed in [43] to mitigate PUEA in cognitive wireless sensor networks. A study about various types of attacks and their countermeasures in wireless sensor networks has been made in [44].

In [45], Fenton's approximation and Wald's sequential probability ratio test (WSPRT)-based scheme has been proposed to mitigate PUEA. Probability of missing was the main parameter considered to set the threshold value. Modified combinational identification algorithm has been discussed in [46] to mitigate the attacks in cooperative sensing. Cluster-based technique to overcome the rogue signal intrusion in cooperative spectrum sensing has been discussed in [47]. In [48], a novel method has been suggested to mitigate the threat in cooperative spectrum sensing. It includes two phases namely identifying phase and sensing phase. In the identifying phase, reliable SUs are found and the sensing results are collected in the second phase. In [49], a trustworthy cognitive radio network has been suggested to defend against malicious users. It is based on the trust value generated and distributed among the nodes. In [50], two algorithms are derived namely encryption algorithm and displacement algorithm from overcoming PUEA. Adaptive orthogonal matching pursuit algorithm (AOMP) has been proposed in [51] to mitigate PUEA. Energy detection, cylostationary and neural network-based scheme have been reported in [52] to cancel PUEA. AND/OR rule-based sensing method has been suggested in [53] to mitigate in PUEA in cooperative spectrum sensing. Improvements in the probability of error is obtained by the OR rule than the AND rule. Nash equilibrium-based differential game method has been suggested in [54] to mitigate PUEA. A new cooperative spectrum sensing in the presence of PUEA has been offered in [55]. Based on the channel information among PU, SU and attackers, weights are derived for optimal combining in the fusion center. A hybrid defence scheme against PUEA with motional secondary users was discussed in [56]. A new spectrum decision protocol to mitigate PUEA in dynamic access networks has been discussed in [57].

1.3.1. Other attacks

Some of the other attacks in the physical layer are denial of service (DOS) attack and replay attack. Any attack in the path between cognitive radio and primary user is known as DOS attack. The malicious user eavesdrop some primary user information and transmit to the cognitive radio at an irrelevant time. This confuses the cognitive radio in deciding the existence of the primary user. This attack is termed as replay attack.

A study about denial of service attack has been made in [58, 59]. Radio frequency fingerprint-based technique has been suggested in [60] to combat DOS attack. Dynamic and smart spectrum sensing algorithm (DS3) has been generated in [61] to minimise the DOS attack. Around 90% of improvement in spectrum utilisation was obtained with the inclusion of DS3 algorithm. Channel eviction triggering scheme in the presence of Rayleigh fading channel has been proposed in [62] to mitigate DOS attack in cooperative spectrum sensing environment. This mechanism is aimed at reducing the misreports and increasing the trustworthy score. A study about replay attack in cognitive radio has been made in [18, 63–65]. A study about the malicious activities in ZigBee network has been made in [66].

1.4. Performance metrics

Performance metrics are used to analyse the system's behaviour and performance. They are used to confirm and validate the specified system performance requirements and to identify the performance issues in a given system.

The important performance metrics for cognitive radio are

- Probability of detection (P_d): Probability of detection is the time during which the primary user is detected.
- Probability of false alarm (P_f): the erroneous detection of the primary user
- Probability of missed detection: failing to detect the primary user. Probability of false alarm: A study about the performance metric has been made in [67–69].
- Receiver operating characteristics (ROC): It is the graph plotted between sensitivity and false positive rate. Here, it is plotted between probability of missed detection and probability of false alarm.

This chapter gives a brief idea about the working of frequency domain-based energy detection spectrum sensing scheme and provides a solution to mitigate PUEA through the authentication tag generated by the collaborator cognitive radio. The sample graphs are plotted between probability of detection and signal to noise ratio, P_d versus P_f.

2. Method to mitigate PUEA

2.1. Collaborator node

To ensure proper spectrum sensing, cognitive radio does not carry out spectrum sensing of its own. Instead, it depends on the third party called collaborator node. It is assumed that the collaborator node is very close to the primary user. The purpose of choosing collaborator node is due to Federal Communication Commissions (FCC) decision 'no modifications must be done to the primary user signal'.

The sample graph is shown in **Figure 4**. The collaborator node senses the availability of the primary user and in the absence of the primary user conveys the message to the cognitive

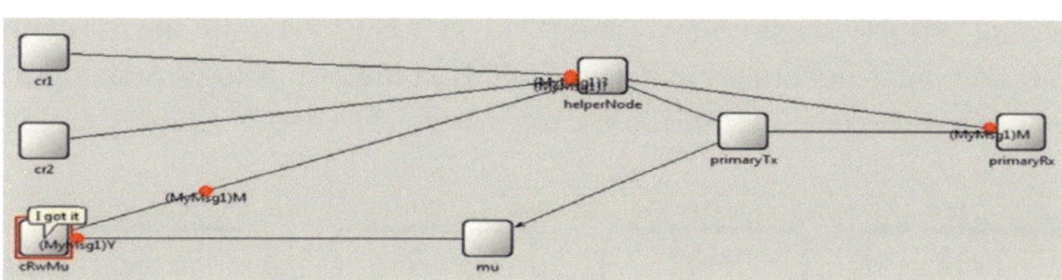

Figure 4. PUEA mitigation.

radio along with the authentication tag. To elude interference with the primary user, the collaborator node communicates with the cognitive radio only in the absence of the primary user. The key to decode the authentication tag is already known to the cognitive radio. The cognitive radio accepts the information only with authentication tag and discards other information. By this way, PUEA is mitigated.

2.2. Spectrum sensing

The collaborator node senses the availability of the primary user with the aid of energy detection method. The block diagram of frequency domain-based energy detection method is shown in **Figure 5**. The incoming signal is filtered and passed to fast Fourier transform block. The output of FFT block is fed to windowing function block. This is done so to reduce the irregularities and to reduce the side lobes. Various windows like Hanning window, Hamming window, Blackman window and Kaiser window could be utilised. Every window has its own advantage and disadvantage. By adjusting beta parameter of Kaiser window, side lobes can be reduced when compared to other windows; but at the same time, the width of main lobe is wider. By adjusting the size of the windows, better output could be obtained. Hence, proper choice of window becomes necessary. The output of windowing block is fed to magnitude square block. The average energy of the signal is then compared with the decision threshold [70–73].

If the incoming signal falls below the threshold, it is null hypothesis (H_0). Only noise is present in the channel and the primary user signal is absent. The spectrum is vacant and could be utilised by the cognitive radio. On the other hand, if the incoming signal exceeds the threshold the decision made is 'primary user present'.

Table 1 summarises the simulation parameters of the graph plotted below. **Figure 6** shows the sample result plotted between P_d versus SNR. SNR is considered as x-axis and P_d as y-axis. For the probability of detection of 0.9, the SNR is −14 dB. The negative scale indicates that the cognitive radio can pick up the primary user signal in a week SNR environment.

Figure 7 shows the output of energy detector for different values of SNR with AWGN noise present in the channel. From the figure, it is clear that as the SNR increases error reduces. Probability of missed detection is lesser for SNR of −5 dB when compared to −20 dB. Lesser the SNR, more is the noise which makes it difficult to detect the presence of the primary user.

2.3. Authentication tag generation by the collaborator node

Once the sensing process is complete, the second step is to generate the authentication tag. The authentication tag is generated in three ways. First method is logic map algorithm-based sequence generation. Second method is by means of DNA-based cryptographic algorithm

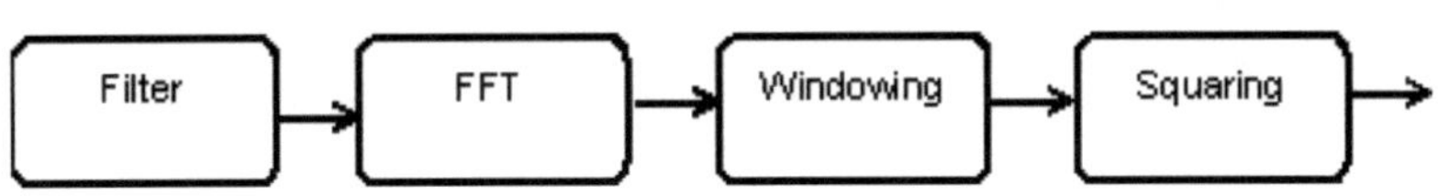

Figure 5. Energy detection method.

Number of samples	300
Probability of false alarm	0.1
Window function	Hanning
Channel	AWGN
FFT size	128

Table 1. Simulation parameters.

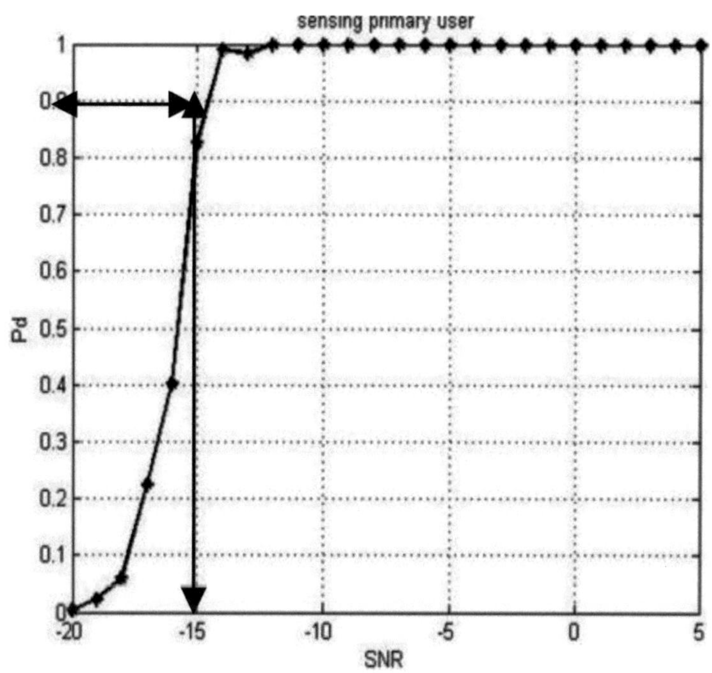

Figure 6. Spectrum sensing.

the sequence is generated. Third method is based on gold code. Utilising gold code generator gold codes are generated. In this, the initial seed value for the gold code is the sequences obtained from the first two methods. The final output from the gold code is treated as the authentication tag to mitigate PUEA.

2.3.1. *Chaotic sequence*

Chaotic sequences help to retrieve the data from intruder in many ways:

a. It changes the transmitted signal into unwanted noise, and therefore it will provide great confusion to the intruder.

b. Code sequences will not repeat for each and every bit of information so it causes the malicious user to take long time to find the sequences.

c. Developing chaotic sequence is simple for both transmitter and receiver who knows the data and parameters used in that transmission, the exact regeneration of data is difficult for a receiver those who wrongly estimate the value. A slight deviation in estimation leads to increasing the error. This is because of sensitivity of chaotic system on their initial condition.

2.3.1.1. Logistic chaotic sequence

1-D logistic chaotic sequence is widely used in communication because of their fast computation process, and simple nature.

Logistic chaotic sequence can be generated by using an expression

$$x(j+1) = r \times x(j) \times (1 - x(j)) \tag{6}$$

where r is called as control parameter and constant, it ranges from $3.57 < r < 4$, $x(1) = 0.99$.

One of the main properties of this sequence is extreme sensitivity to initial condition and good correlation property.

Figure 8 shows the signal to noise ratio versus primary user detection graph plotted with and without authentication tag. The overlapping of both the graphs shows that there is no significant change in the performance of the collaborator system when an authentication tag is inserted. The authentication tag and the spectrum-free information are transmitted to the cognitive radio. The probability of false alarm is fixed as 0.1 and the number of samples chosen is 300. Additive white Gaussian noise (AWGN) is considered as the channel noise.

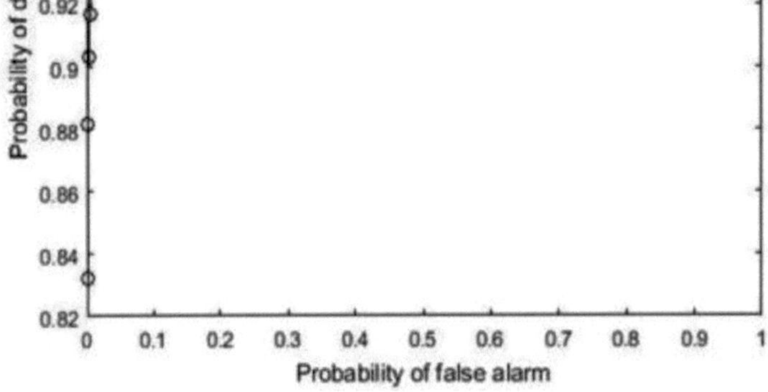

Figure 7. Comparison between various SNR.

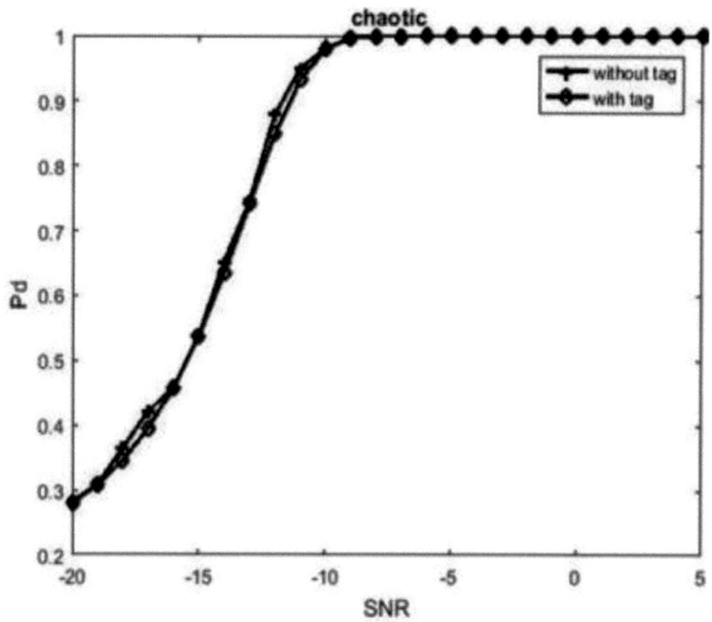

Figure 8. Chaotic-based tag generation.

2.3.2. DNA

DNA algorithm has been utilised in this work to generate the authentication tag because the storage and processing of data is very secure. One single DNA can be split into four basic units. They are Adenine (A), Thymine (T), Cytosine (C) and Guanine (G). So, it is also known as quaternary encoding. Binary values are assigned to these units for encoding purpose as follows:

A—00, T—01, C—10 and G—11.

Algorithm

Step 1: Transform message bits into binary

Step 2: Assign A, T, G and C to binary(a)

Step 3: Get key value from server(b)

Step 4: Take one's complement to step 2 and 3

Step 5: Do XOR operation between output from step 4(a′ and b′)

Step 6: Transform bits from step 5 into DNA form

Step 7: Transform DNA form into ASCII values

Step 8: Transform into binary form(encrypted)

Figure 9 shows the signal to noise ratio versus probability of detection graph plotted with and without authentication tag. The overlapping of both the graphs shows that there is no notable

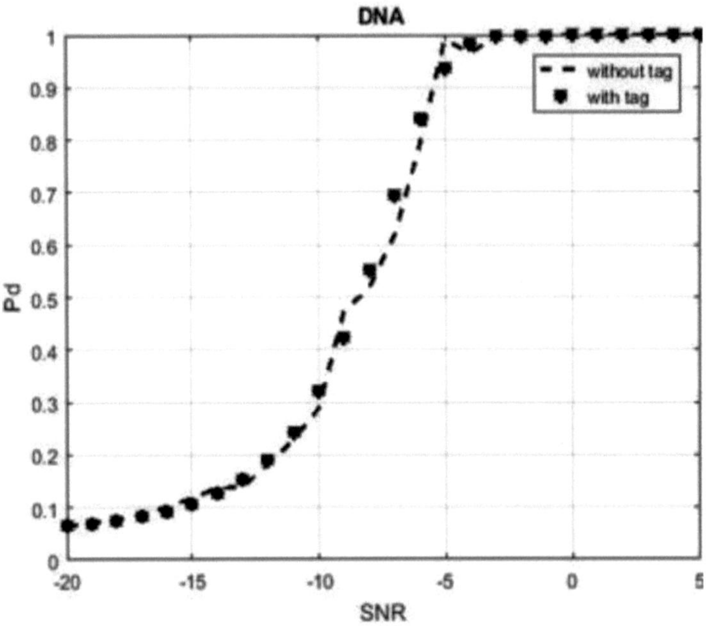

Figure 9. DNA algorithm-based tag generation.

difference in the performance of the collaborator system when an authentication tag is added along with the primary user availability information.

2.3.3. Gold code

Pseudonoise (PN) is a signal similar to noise but generated with a definite pattern. In cryptography, PN sequences are widely to ensure data protection from intruders. The PN sequences are added with the message signal so that it appears as noise to the malicious users. Various types of PN sequences are available. Their auto- and cross-correlation properties decide the choice of PN sequences. Some PN sequences have good autocorrelation property but not cross-correlation property. Some have good cross-correlation property but not autocorrelation property. Gold code is chosen because of its good auto and cross-correlation property. Gold codes are obtained by mod-2 addition of shifted pairs of m-sequences with length m. The autocorrelation and cross-correlation function of gold code, $2^t - 1$, is

Autocorrelation function:

$$\varphi_{GC}(h)$$

$$\text{Where } \varphi_{GC}(h) = \{\pm 2^t - 1, h = 0 \qquad (7)$$

$$\pm 1, h \neq 0$$

http://dx.doi.org/10.5772/intechopen.80328

Cross-correlation function:

$$\psi_{GC}(h)$$
$$\text{Where } \psi_{GC}(h) = (2^t - 1, h = \lambda \tag{8}$$
$$\pm 1, h \neq \lambda$$

2.3.3.1. *Trilayered authentication*

The proposed work is to integrate all the three algorithms and to generate a trilayered authentication tag to mitigate PUEA. Both the LFSRs required a seed value for their functioning. Hence, the initial seed value of one LFSR is the sequence generated utilising DNA algorithm and for the second LFSR it is a chaotic sequence. The outputs from the LFSRs are XORed, and the resulting gold code sequence is considered an authentication tag. It is as shown in **Figure 10**.

Figure 11 shows the sample signal to noise ratio versus probability of detection graph plotted with and without authentication tag. From the figure, it can be depicted that there is no drastic change in the performance of the collaborator system when an authentication tag is add along with the primary user availability information.

Figure 11b shows the graph plotted by increasing the size of the window function. Here, Hamming window of size 10 has been utilised.

Figure 11c shows the plot of signal to noise ratio versus probability of detection graph plotted with and without authentication tag. Here, the FFT size of the energy detector has been raised from 64 to 128.

Figure 11d shows the graph plotted with the probability of false alarm fixed as 0.01.

2.3.3.2. *Hardware implementation*

Universal software-defined radio peripheral (USRP) is a universally accepted test bed for cognitive radio. The USRP software-defined radio device is a tuneable transceiver. It is used as a prototype for wireless communication systems. It offers frequency ranges up to 6 GHz with up to 56 MHz of instantaneous bandwidth. It allows advanced wireless applications to be created with LabVIEW, enabling rapid prototyping.

The prototype of energy detection-based spectrum sensing scheme is developed using LabVIEW tool. LabVIEW is a modelling, simulation and real-time implementation tool which

Figure 10. Trilayered authentication.

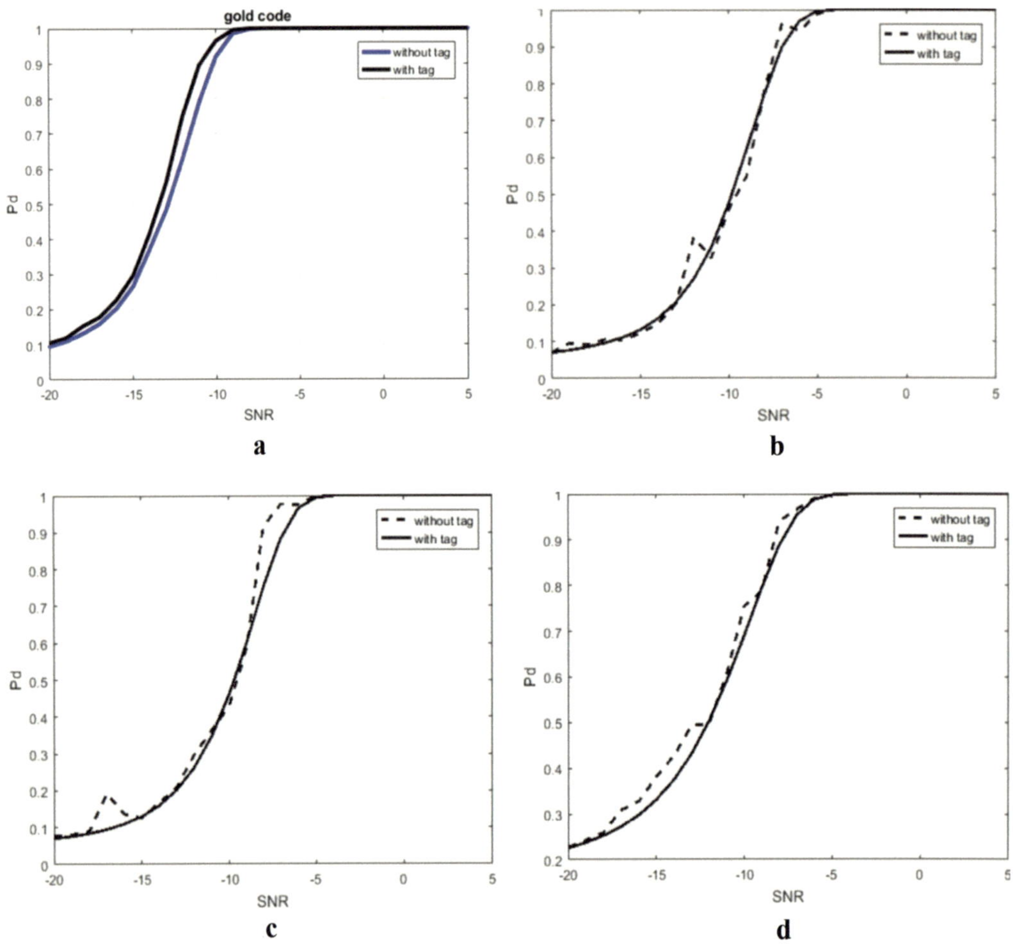

Figure 11. (a)–(d) Trilayer-based tag generation.

is being used around the world for implementation development through software. It uses runtime engine to simulate the designs. Front panel and block panel support the graphical user interface (GUI) structure of LabVIEW. Front panel comprises of controls and indicators, whereas block panel has functions, structures, Sub-Vis and terminals to execute the required design.

The transmitter and the receiver blocks are developed using LabVIEW software. **Figure 12** shows the block diagram of energy detector. Once the blocks are developed using LabVIEW software then the physical connections are made. Ethernet cable is used to connect USRP with the computer in which the blocks are developed.

Then, the signal is transmitted using USRP. **Figure 13** shows the USRP front panel.

Figure 14 shows the experimental setup using USPR. Out of two USRPs, one USRP is treated as transmitter and the other USRP is treated as receiver. Additive white Gaussian noise (AWGN) is considered as the noise in the channel.

http://dx.doi.org/10.5772/intechopen.80328

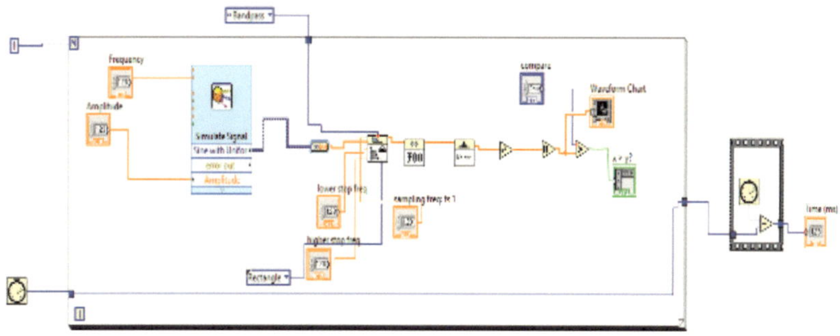

Figure 12. LabVIEW-based energy detector.

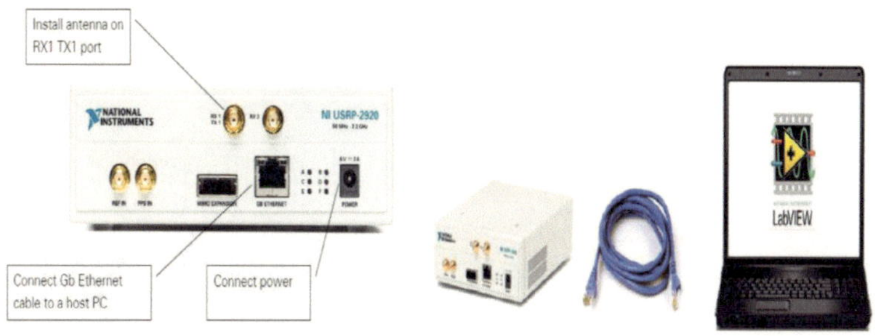

Figure 13. Front panel of USRP.

Figure 14. USRP experimental setup.

Table 2 shows the specification of USRP. For transmission, the IP address is 192.168.10.1 and for reception the IP address is set as 192.168.10.2. The USRPs are connected to the computer via Ethernet cable. The distance between the two USRPs is set as 100 cm.

Figure 15a shows the transmission of primary user signal at the transiting end and **Figure 15b** shows the detection of primary user signal at the receiving end. The received signal is now compared with the threshold value. The incoming signal exceeds the threshold value. The presence of primary user is detected and plotted. For an SNR of –5 dB, the probability of detection is 0.9.

Frequency range	50 MHz–2.2 GHz
Gain range	0–31 dB
Frequency accuracy	2.5 ppm
DAC	2 channels, 16 bit
Noise figure	5–7 dB
Maximum I/Q sampling rate	16-bit sample width at 20 MHz, 8-bit sample width at 40 MHz

Table 2. USRP specifications.

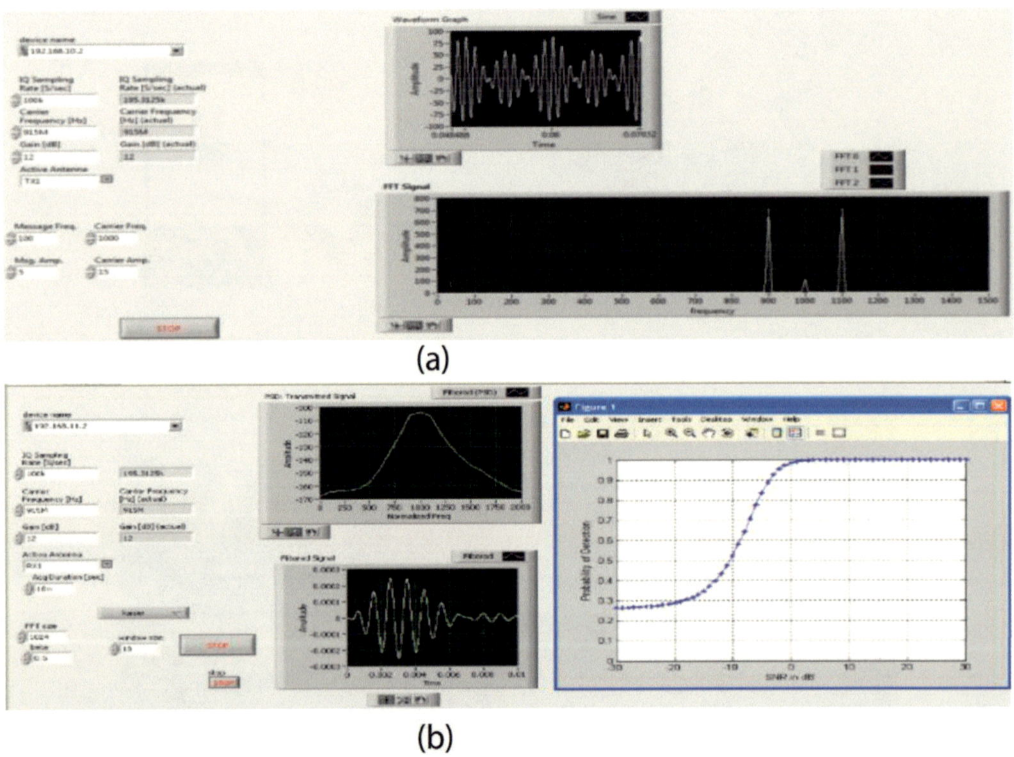

Figure 15. (a) Transmission using USRP. (b) Reception using USRP.

3. Conclusion

To avoid wastage of bandwidth and to achieve dynamic spectrum access cognitive radio is the best solution. To achieve dynamic spectrum access, the most important function of cognitive radio is spectrum sensing.

In this chapter,

- Energy detection-based spectrum sensing scheme has been discussed to detect the existence of the primary user by the collaborator node. This method has been chosen because of its simple nature.
- To combat PUEA, a collaborator node-based approach has been suggested. The cognitive radio requests the collaborator node to sense the free spectrum. The collaborator node senses the availability of the primary user.
- Once the availability of the free spectrum is confirmed, the message has been conveyed to the cognitive radio in a secure manner. Hence, a trilayered method has been suggested to generate the authentication tag. The message along with the tag is accepted by the CR and others are rejected. By this way, the PUEA attack has been overruled. Threat-free environment makes the cognitive radio to arrive at a proper conclusion about the presence of spectrum holes and utilise it.

Acknowledgements

The Authors would like to express their sincere thanks to SASTRA Deemed University, for the grant received under R&M fund (R&M/0027/SEEE – 010/2012–13) to carry out this book chapter.

Author details

Avila Jayapalan* and Thenmozhi Karuppasamy

*Address all correspondence to: avila@ece.sastra.edu

SASTRA Deemed to be University, Thanjavur, Tamil Nadu, India

References

[1] Mitola J, Maguire GQ. Cognitive radio: Making software radios more personal. IEEE Personal Communications. 1999;**64**:13-18

[2] Haykin S. Cognitive radio: Brain empowered wireless communications. IEEE Journal on Selected Areas in Communications. 2005;**232**:201-220

[3] Igbinosa IE, Oyerinde OO, Srivastava VM, Mneney S. Spectrum sensing methodologies for cognitive radio systems: A review. International Journal of Advanced Computer Science and Applications. 2015;**6**(12):13-22

[4] Akyildiz IF, Lee WY, Vuran MC, Mohanty S. A survey on spectrum management in cognitive radio networks. IEEE Communications Magazine. 2008;**46**(4):40-48

[5] Kaur M, Kaur A. Cognitive radio spectrum sharing techniques: A review. International Journal of Computer Science and Information Technologies. 2015;**6**(3):3089-3091

[6] Asokan A, Ayyappadas R. Survey on cognitive radio and cognitive radio sensor networks. In: International Conference on Electronics and Communication Systems; 2014. pp. 1-7

[7] Furtado A, Irio L, Oliveira R, Bernardo L, Dinis R. Spectrum sensing performance in cognitive radio networks with multiple primary users. IEEE Transactions on Vehicular Technology. 2015;**PP**(99):1-11

[8] Jia J, Zhang Q, Shen XS. HC-MAC: A hardware-constrained cognitive MAC for efficient spectrum management. IEEE Journal on Selected Areas in Communications. 2008; **26**(1):106-117

[9] Sabuj SR, Hamamura M, Kuwamura S. Detection of intelligent malicious user in cognitive radio network by using Friend or Foe (FoF) detection technique. In: International Telecommunication Networks and Conference; 2015. pp. 155-160

[10] Muchandi N, Khanai R. Cognitive radio spectrum sensing: A survey. In: International Conference on Electrical, Electronics, and Optimization Techniques (ICEEOT); 2016. pp. 3233-3237

[11] Anyim IG, Chiverton J, Filip M, Tawfik A. The optimization of wideband cyclostationary feature detector with receiver constraints. In: IET 3rd International Conference on Intelligent Signal Processing (ISP 2017); 2017. pp. 1-6

[12] Kakalou I, Papadopoulou D, Xifilidis T, Psannis KE, Siakavara K, Ishibashi Y. A survey on spectrum sensing algorithms for cognitive radio networks. In: 7th International Conference on Modern Circuits and Systems Technologies (MOCAST); 2018. pp. 1-4

[13] Orumwense E, Oyerinde O, Mneney S. Improving trustworthiness amongst nodes in cognitive radio networks. In: Southern Africa Telecommunication Networks and Applications Conference; 2014. pp. 401-406

[14] Jin Z, Anand S, Subbalakshmi KP. Robust spectrum decision protocol against primary user emulation attacks in dynamic spectrum access networks. In: IEEE Global Telecommunications Conference; 2010. pp. 1-5

[15] Jain S, Dhawan A, Jha CK. Emulation attack in cognitive radio networks: A study. International Journal of Computer Networks and Wireless Communications (IJCNWC). 2014;**4**(2):169-172

[16] Lakshmibai T, Chandrasekaran B, Parthasarathy C. Primary user authentication in cognitive radio networks: A survey. International Journal of Advanced Research in Electrical, Electronics and Instrumentation Engineering. 2014;**3**(1):6935-6942

[17] Tang L, Wu J. Research and analysis on cognitive radio network security. Wireless Sensor Network. 2012;**4**:120-126

[18] Alhakami W, Mansour A, Safdar GA. Spectrum sharing security and attacks in CRNs: A review. International Journal of Advanced Computer Science and Applications. 2014;**5**(1):76-87

[19] Tabatabaee S, Bagheri A, Shahini A, Shahzadi A. An analytical model for primary user emulation attacks in IEEE 802.22 networks. In: International Conference on Connected Vehicles and Expo; 2013. pp. 693-698

[20] Kim H. End-to-end authentication protocols for personal/portable devices over cognitive radio networks. International Journal of Security and Its Applications. 2014;**8**(4): 123-138

[21] Shan-shan W, Xing-guo L, Bai-nan L. Primary user emulation attacks analysis for cognitive radio networks communication. TELKOMNIKA. 2013;**11**(7):3905-3914

[22] Xiea X, Wang W. Detecting primary user emulation attacks in cognitive radio networks via physical layer network coding. Journal of Ubiquitous Systems & Pervasive Networks. 2014;**5**(1):1-8

[23] Das D, Das S. Eigenvalue detection based method to mitigate PUEA in cognitive radio networks. IEEE International Conference on Advanced Networks and Telecommunication Systems; 2013. pp. 1-6

[24] Liu Y, Ning P. Enhanced wireless channel authentication using time-synched link signature. In: Proceedings IEEE INFOCOM; 2012. pp. 2636-2640

[25] Patwari N, Kasera SK. Robust location distinction using temporal link signatures. IEEE Transactions on Mobile Computing. 2010;**10**(3):449-462

[26] Liu Y, Ning P, Dai H. Authenticating primary users' signals in cognitive radio networks via integrated cryptographic and wireless link signatures. In: IEEE Symposium on Security and Privacy; 2010. pp. 286-301

[27] Wang W, Li H, Sun Y, Han Z, Attack-proof collaborative spectrum sensing in cognitive radio networks. In: 43rd Annual Conference on Information Sciences and Systems; 2009. pp. 130-134

[28] Salem F, Ibrahim MH, El-Wahab Ali IA. Secure authentication scheme preventing wormhole attacks in cognitive radio networks. Asian Journal of Computer Science and Technology. 2013;**2**(4):52-55

[29] Salem FM, Ibrahim MH, Ibrahim II. A primary user authentication scheme for secure cognitive TV spectrum sharing. International Journal of Computer Science Issues. 2012;**9**(4): 157-166

[30] Saber MJ, Sadough SMS. Optimal energy detection in cognitive radio networks in the presence of malicious users, In: Third International Conference on Computer and Knowledge Engineering; 2013. pp. 173-177

[31] Alahmadi A, Abdelhakim M, Ren J, Li T. Mitigating primary user emulation attacks in cognitive radio networks using advanced encryption standard. In: IEEE Global Communications Conference; 2013. pp. 3329-3334

[32] Borle KM, Chen B, Du W. A physical layer authentication scheme for countering primary user emulation attack. In: IEEE International Conference on Acoustics, Speech and Signal processing; 2013. pp. 2935-2939

[33] Chandrashekar S, Lazos L. A primary user authentication system for mobile cognitive radio networks. In: 3rd International Symposium on Applied Sciences in Biomedical and Communication Technologies; 2010. pp. 1-5

[34] Gao Z, Zhu H, Li S, Suguo D, Xu L. Security and privacy of collaborative spectrum sensing in cognitive radio networks. IEEE Wireless Communications. 2012;**19**(6):1536-1584

[35] Chen R, Park J-M, Reed JH. Defense against primary user emulation attacks in cognitive radio networks. IEEE Journal on Selected Areas in Communications. 2008;**26**(1):25-37

[36] Di P, Alexander M. Wyglinski, primary-user emulation detection using database-assisted frequency-domain action recognition. IEEE Transactions on Vehicular Technology. 2014;**63**(9):4372-4382

[37] Jo M, Han L, Kim D, In HP. Selfish attacks and detection in cognitive radio ad-hoc networks. IEEE Network. 2013;**27**(3):46-50

[38] Sujitha R, Poornima. Efficient detection of selfish attacks in cognitive radio networks using COOPON algorithm. International Journal of Advanced Research Trends in Engineering and Technology. 2015;**2**(23):84-91

[39] Saletm FM, Ibrahim MH, Ali IA, Ibrahim II. Matched-filter-based spectrum sensing for secure cognitive radio network communications. International Journal of Computer Applications. 2014;**87**(18):41-44

[40] Nomura R, Masahiro K, Mizuno T. Radio-free mutual authentication for cognitive radio network. Information and Media Technologies. 2011;**6**(2):580-594

[41] Kim HS. Location-based authentication protocol for first cognitive radio networking standard. Journal of Network and Computer Applications. 2011;**3**(4):1161-1167

[42] Taheri S, Sharifi A, Berangi R. Deal with attacks over cognitive radio networks authentication. International Journal of Computer Technology & Applications; 2012;**2013**(6): 2027-2032

[43] Blesa J, Romero E, Rozas A, Araujo A. PUE attack detection in CWSNs using anomaly detection techniques. EURASIP Journal on Wireless Communications and Networking. 2013:1-13

[44] Araujo A, Blesa J, Romero E, Villanueva D. Security in cognitive wireless sensor networks. Challenges and open problems. EURASIP Journal on Wireless Communications and Networking. 2012;**2012**:1-8

[45] Jin Z, Anand S, Subbalakshmi KP. Detecting primary user emulation attacks in dynamic spectrum access networks. In: IEEE International Conference on Communications; 2009. pp. 1-5

[46] Qi Z, Li Q, Hsieh G. Against cooperative attacks in cooperative spectrum sensing. IEEE Transactions on Wireless Communications. 2013;**12**(6):2680-2687

[47] Jackson DS, Zang W, Gu Q, Cheng W, Yu M. Exploiting and defending trust models in cooperative spectrum sensing. EURASIP Journal on Wireless Communications and Networking. 2015;**2015**:1-18

[48] Lu J, Ping Wei A. Scheme to counter SSDF attacks based on hard decision in cognitive radio networks. WSEAS Transactions on Communications. 2014;**13**:242-248

[49] Orumwense E, Oyerinde O, Mneney S. Improving trustworthiness amongst nodes in cognitive radio networks. In: Conference on South Africa Telecommunication Networks and Applications; 2014. pp. 401-406

[50] Zhou X, Xiao Y, Li Y. Encryption and displacement based scheme of defense against primary user emulation attack. In: 4th IET International Conference on Wireless, Mobile & Multimedia Networks; 2011. pp. 44-49

[51] Dang M, Zhao Z, Zhang H. Detection of primary user emulation attacks based on compressive sensing in cognitive radio networks. In: International Conference on Wireless Communications & Signal Processing; 2013. pp. 1-5

[52] Pu D, Shi Y, Ilyashenko AV, Wyglinski AM. Detecting primary user emulation attack in cognitive radio networks. In: IEEE Global Telecommunications Conference; 2011. pp. 1-5

[53] Haghighat M, Sadough SM, Cooperative spectrum sensing in cognitive radio networks under primary user emulation attacks. In: 6th International Symposium on Tele communications; 2012. pp. 148-151

[54] Hao D, Sakurai K. A differential game approach to mitigating primary user emulation attacks in cognitive radio network. In: 26th IEEE International Conference on Advanced Information Networking and Applications; 2012. pp. 495-502

[55] Chen C, Cheng H, Yao Y-D. Cooperative spectrum sensing in cognitive radio networks in the presence of the primary user emulation attack. IEEE Transactions on Wireless Communications. 2011;**10**(7):2135-2141

[56] Bao F, Chen H, Xie L. Analysis of primary user emulation attack with motional secondary users in cognitive radio networks. In: IEEE 23rd International Symposium Personal Indoor and Mobile Radio Communications; 2012. pp. 956-961

[57] Jin Z, Anand S, Subbalakshmi KP. Impact of primary user emulation attacks on dynamic spectrum access networks. IEEE Transactions on Communications. 2012;**60**(9):2635-2643

[58] Brown TX, Sethi A. Potential cognitive radio denial-of-service vulnerabilities and protection countermeasures: A multi-dimensional analysis and assessment, mobile network applications. In: 2nd International Conference on Cognitive Radio Oriented Wireless Networks and Communications; 2008. pp. 456-464

[59] Wang W. Denial of service attacks in cognitive radio networks. In: 2nd Conference on Environmental Science and Information Application Technology; 2010. pp. 530-533

[60] Soderi S, Dainelli G, Hämäläinen M, Iinatti J. Signal fingerprinting in cognitive wireless networks. In: International Conference on Cognitive Radio Oriented Wireless Networks; 2014. pp. 267-270

[61] Amjad MF, Aslam B, Zou CC. DS3: A dynamic and smart spectrum sensing technique for cognitive radio networks under denial of service attack. In: IEEE Global Communications Conference; 2013. pp. 1149-1154

[62] Sodagari S, Attar A, Leung VCM, Bilen SG. Denial of service attacks in cognitive radio networks through channel eviction triggering. In: IEEE Global Telecommunications Conference; 2010. pp. 1-5

[63] Butt MA, Zaman M. Cognitive radio network: Security enhancements. Journal of Global Research in Computer Science. 2013;**2**(4):36-41

[64] Idoudi H, Daimi K, Saed M. Security challenges in cognitive radio networks. In: Proceedings of the World Congress on Engineering; 2014. pp. 498-504

[65] Sen J. A survey on security and privacy protocols for cognitive wireless sensor networks. Journal of Network and Information Security. 2013;**1**(1):1-31

[66] Mangir T, Sarakbi L, Younan H. Detecting malicious activities in ZigBee networks using cognitive radio. International Journal of Distributed and Parallel Systems. 2011;**2**(6):51-62

[67] Sood V, Singh M. On the performance of detection based spectrum sensing for cognitive radio. International Journal of Electronics & Communication Technology. 2011;**2**(3):140-143

[68] Pal RU, Indurkar PR, Lakhe PR. Review of performance of energy detection, matched filter detection and cyclostationary detection based spectrum sensing under different wireless channels. International Journal of Engine Research. 2015;**3**(2):6-9

[69] Buttar S. Comparison of energy detection in cognitive radio over different fading channels. International Journal of Advancements in Research and Technology. 2012;**1**(2):99-105

[70] Kim YM, Zheng G, Sohn SH, Kim JM. An alternative energy detection using sliding window for cognitive radio system. In: Tenth International Conference on Advanced Communication Technology; 2008. pp. 481-485

[71] http://www.ni.com/white-paper/4844/en/

[72] Khatun M. Implement a new window function and design FIR filters by using this new window. International Journal of Engineering and Computer Science. 2014;**3**(3):4087-4090

[73] Wang H, Noh G, Kim D, Kim S, Hong D. Advanced sensing techniques of energy detection in cognitive radios. Journal of Communications and Networks. 2010;**12**(1):19-29

Chapter 6

Interference Alignment in Multi-Input Multi-Output Cognitive Radio-Based Network

Arif Basgumus, Mustafa Namdar, Hakan Alakoca,
Eylem Erdogan and Lutfiye Durak-Ata

Additional information is available at the end of the chapter

http://dx.doi.org/10.5772/intechopen.80073

Abstract

This study investigates the interference alignment techniques for cognitive radio networks toward 5G to meet the demand and challenges for future wireless communications requirements. In this context, we examine the performance of the interference alignment in two parts. In the first part of this chapter, a multi-input multi-output (MIMO) cognitive radio network in the presence of multiple secondary users (SUs) is investigated. The proposed model assumes that linear interference alignment is used at the primary system to lessen the interference between primary and secondary networks. Herein, we derive the closed-form mathematical equations for the outage probability considering the interference leakage occurred in the primary system. The second part of this study analyzes the performance of interference alignment for underlay cognitive two-way relay networks with channel state information (CSI) quantization error. Here, a two-way amplify-and-forward relaying scheme is considered for independent and identically distributed Rayleigh fading channel. The closed-form average pairwise error probability expressions are derived, and the effect of CSI quantization error is analyzed based on the bit error rate performance. Finally, we evaluate the instantaneous capacity for both primary and secondary networks*.

Keywords: 5G wireless communication systems, average pairwise error probability, CSI quantization, cognitive radio networks, interference alignment, MIMO, outage probability performance, two-way amplify-and-forward relaying

*The content of this study has partially been submitted in IEEE 41st International Conference on Telecommunications and Signal Processing (TSP 2018).

1. Introduction

The rapidly growing number of mobile devices, higher data rates and cellular traffic, and quality of service requirements trigger the development of mobile communications. It is expected that the next-generation cellular networks (5G and beyond) will meet the advanced technology requirements. 4G networks are not powerful enough to support massively connected devices with low latency and high spectral efficiency, which is critical for next-generation networks. 5G networks are characterized by three fundamental functions in general: connectivity for everywhere, low latency for communication, and very high-speed data transmission [1].

In the near future, a large number of mobile devices will connect to one another in everywhere and provide a seamless mobile user experience. Real-time applications and critical systems and services (medical applications, traffic flow, etc.) with zero latency are expected to be offered over 5G cellular networks. Besides, the fast data transmission and reception will be ensured by supporting zero latency using a high-speed link. For this reason, the scope of 5G cellular networks bring the emerging advantages, new architectures, methodologies, and technologies on telecommunications such as energy-efficient heterogeneous networks, software-defined networks (SDN), full-duplex radio communications, device-to-device (D2D) communications, and cognitive radio (CR) networks. An increasing number of mobile devices and the bandwidth requirement for large amounts of data require the development of the new technologies and infrastructures in addition to the existing technology. It is inevitable that the number of smart phones, high-definition televisions, cameras, computers, transport systems, video surveillance systems, robots, sensors, and wearable devices produces a huge amount of voice-data traffic in the near future. To meet the growth and to provide fast and ubiquitous Internet access, several promising technologies have been developed. Regarding the deployment of the 5G wireless communication systems, the corresponding growth in the demand for wireless radio spectrum resources will appear. The capacity of the communication networks will be increased by using the energy-efficiency techniques with the evolving technology in 5G networks [2–5].

One of the candidates for solving the problem of spectrum shortage is the CR network which will be a key technology for 5G networks. CR has attracted considerable interest as it can cope with the spectrum underutilization phenomenon. Performing spectrum sharing using a CR network is an important issue in wireless communication networks. There are three main ways for a primary network user to share the frequency spectrum with a cognitive user: underlay, overlay, and interweave. In the underlay method, the secondary user (SU) transmits its information simultaneously with the primary user (PU) as long as the interference between SU and PU receivers is within a predefined threshold. In the overlay approach, SU helps PU by sharing its resources, and in return, PU allows SU to communicate. In the interweave technique, SU can use the bandwidth of PU if PU is not active. In this model, SU should have perfect spectrum-sensing features to analyze the spectrum [6–9].

Among the various methods of solving the interference problem, interference alignment (IA) is one of the most promising ways to achieve it. IA is an important approach for CR to recover the desired signal by utilizing the precoding and linear suppression matrices which consolidates the interference beam into one subspace in order to eliminate it [10–13]. In the literature, linear IA is adopted in CR interference channels in [14–20] and the references therein. In [14], adaptive power allocation schemes are considered for linear IA-based CR networks where the outage probability and sum rate were derived. In [15], adaptive power allocation was studied for linear IA-based CR using antenna selection at the receiver side. Ref. [16] enhances the security of CR networks by using a zero-forcing precoder. Moreover, in [17], a similar work was proposed to improve the overall outage performance of the interference channel by using power allocation optimization. These studies have shown that interference management is a critical issue to be handled in all multiuser wireless networks.

CR technology can be capable of utilizing the spectrum efficiently as long as the interference between PU and SU is perfectly aligned as shown in **Figure 1**. A set of studies discussing IA is presented in the literature [21–29].

Motivated by the above works, in the first part of this study, we examine the impact of interference leakage on multi-input multi-output (MIMO) CR networks with multiple SUs. Specifically, a closed-form outage probability expression is derived to provide the performance of the primary system. Then, in the second part of our work, we investigate the performance of IA in underlay CR networks for Rayleigh fading channel. Moreover, unlike the mentioned papers, the effect of CSI quantization error is taken into account in our analysis. Then, a two-way relaying scheme with amplify-and-forward (AF) strategy is studied. Finally, the effects of the relay location and the path loss exponent on the BER performance and system capacity and CSI quantization on the average pairwise error probability (PEP) performance for this two-way AF system are presented.

The main simulation parameters and their descriptions used in this study are summarized in **Table 1**.

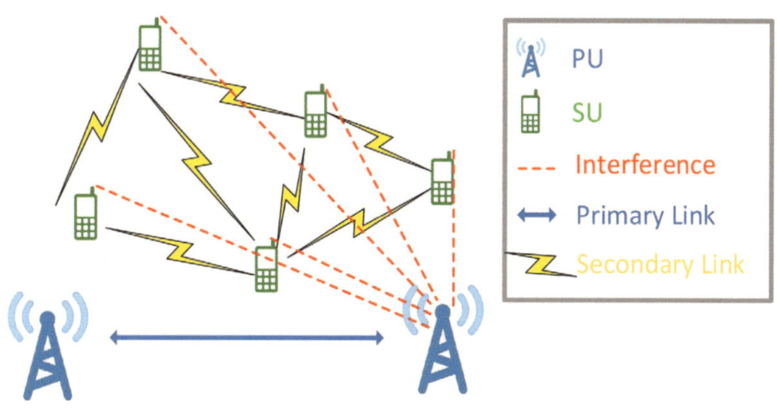

Figure 1. Illustration of the primary link between PU pair and interference links generated by the SUs.

Symbol	Description
P_1 and P_2	Transmitted powers of the PU and SU
σ_N^2	Variance of the circularly symmetric additive white Gaussian noise vector
R_{th}	Data rate threshold
α	Interference-leakage parameter
M_p and N_p	Number of transmit-and-receive antennas of PU
M_s and N_s	Number of transmit-and-receive antennas of SU
K	Number of SU
$d_{j,i}$	Distance between the *i*th transmitter and the *j*th receiver nodes
$\tau_{j,i}$	Path loss exponent between the *i*th transmitter and the *j*th receiver nodes
$B_{j,i}$	Channel state information exchange amount between the *i*th transmitter and the *j*th receiver nodes

Table 1. The simulation symbols and their descriptions.

2. The impact of interference leakage on MIMO CR networks

In this study, MIMO interference alignment-based CR network with a PU and multiple SUs is considered under Rayleigh fading channel.

2.1. System model

In the system model as it is shown in **Figure 2**, the number of transmit-and-receive antennas of the PU is given by M_p and N_p. The transmit antennas at each SU are given as M_s. The received signal, $\mathbf{y}_p$, implementing the IA technique is given as

$$\mathbf{y}_p = \mathbf{U}_p^{\mathrm{H}}\mathbf{H}_{pp}\mathbf{V}_p\mathbf{x}_p + \sqrt{\alpha}\sum_{i=1}^{K}\mathbf{U}_s^{\mathrm{H}}\mathbf{H}_{ps_i}\mathbf{V}_s\mathbf{x}_{s_i} + \mathbf{U}_p^{\mathrm{H}}\mathbf{n}, \tag{1}$$

where x_p and x_{s_i} are the transmitted signals from PU and the *i*th SU (for $i = 1, 2, \ldots, K$), respectively. Herein, $\mathbf{H}_{pp}$ is the matrix of channel coefficients between the PU pair, and $\mathbf{H}_{ps_i}$ denotes the channel matrix between the primary receiver and the *i*th secondary transmitter. The interference leakage is modeled similar to the one in [30]. The interference-leakage parameter α $(0 \leq \alpha \leq 1)$ represents the status of the alignment, i.e., $\alpha = 0$ and 1 corresponds to perfect alignment and perfect misalignment cases, respectively. $\mathbf{V}$ and $\mathbf{U}$ are the precoding- and interference-suppression matrices. The superscript $(\cdot)^{\mathrm{H}}$ denotes the Hermitian operator, and $\mathbf{n}$ is the zero-mean unit variance $(\sigma_N^2 = 1)$ circularly symmetric additive white Gaussian noise (AWGN) vector.

The following conditions must be satisfied for perfect interference alignment between PU and SUs:

http://dx.doi.org/10.5772/intechopen.80073

$$\mathbf{U}_s^{\mathrm{H}}\mathbf{H}_{ps_i}\mathbf{V}_s\mathbf{x}_{s_i} = 0, \tag{2}$$

$$\mathrm{Rank}\left(\mathbf{U}_s^{\mathrm{H}}\mathbf{H}_{ps_i}\mathbf{V}_s\mathbf{x}_{s_i}\right) = d. \tag{3}$$

Each user transmits d data streams. Using the ideal linear IA technique, (1) can be re-expressed as

$$\mathbf{y}_p = \mathbf{U}_p^{\mathrm{H}}\mathbf{H}_{pp}\mathbf{V}_p\mathbf{x}_p + \mathbf{U}_p^{\mathrm{H}}\mathbf{n}. \tag{4}$$

2.2. Outage probability analysis

The channel capacity and outage probability are the most important impairments which affect the quality of service (QoS) in wireless communication systems. When no CSI conditions are

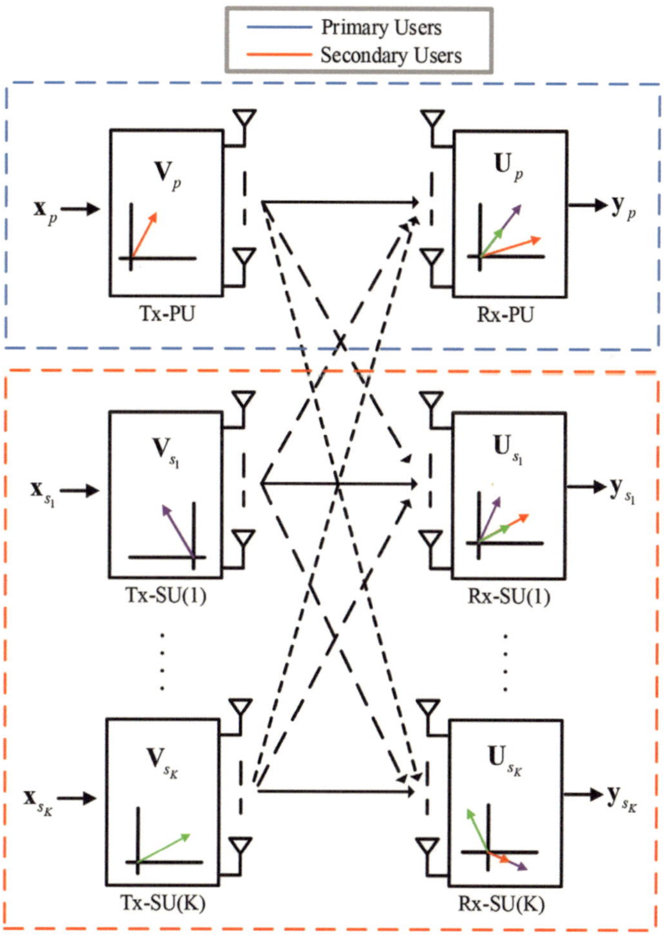

Figure 2. IA-based CR network with single PU and K SUs sharing the spectrum.

given, MIMO channel capacity is expressed as in [31]. The channel capacity of the considered MIMO system in PU can be expressed as

$$C = \log_2 \det \left| \mathbf{I} + \frac{\gamma_1}{(1+\gamma_2)N_p} \mathbf{H}_{pp} \mathbf{H}_{pp}^H \right|, \tag{5}$$

where $\gamma_1 = P_1 \|\mathbf{H}_{pp}\|^2 / \sigma_N^2$ is the signal-to-noise ratio (SNR) of the primary link. γ_2 can be expressed as $\gamma_2 = (P_2/\sigma_N^2) \sum_{i=1}^{K} \|\mathbf{H}_{ps_i}\|^2$. Note that $\|.\|^2$ demonstrates the squared Frobenius norm of the channel matrix, **I** denotes for identity matrix, and P_1 and P_2 are the transmitted powers of the PU and SUs, respectively. If linear IA perfectly eliminates the interference between SU and PU, then SNR of the interference channel, γ_2, becomes zero. It is important to note that precoding and linear suppression vectors are assumed as $\left|\mathbf{U}_p^{\mathrm{H}}\right|^2 = \left|\mathbf{V}_p\right|^2 = \left|\mathbf{U}_{s_i}^{\mathrm{H}}\right|^2 = \left|\mathbf{V}_{s_i}\right|^2 = 1$. In the presence of interference-free communication, primary system works in the single-input and single-output (SISO) fashion [14]. Hence, the probability density function (PDF) of γ_1 can be written as $f_{\gamma_1}(\gamma) = \frac{1}{\overline{\gamma}_1} \exp(-\gamma/\overline{\gamma}_1)$, and the outage probability of the system can be obtained as

$$P_{out} = \int_0^{2^{R_{th}}-1} f_{\gamma_1}(\gamma) d\gamma, \tag{6}$$

where R_{th} is the data rate threshold and $\overline{\gamma}_1 = P_1/\sigma_N^2$ denotes the average SNR of the primary system. By substituting $f_{\gamma_1}(\gamma)$ into (6), the outage probability can be obtained as

$$P_{out} = 1 - \exp\left(\frac{2^{R_{th}} - 1}{\overline{\gamma}_1}\right). \tag{7}$$

In the presence of interference, the primary system works in MIMO fashion, and leakages may occur due to fast-fading Rayleigh channel. To improve the performance of the primary system, we adopt maximum ratio transmission and maximum ratio combining at the transmitter and receiver, respectively. Thereby, the end-to-end signal-to-interference-plus-noise ratio (SINR) of the primary system can be written as $\gamma_\tau = \gamma_1/(1+\gamma_2)$. In the proposed system, all channels are modeled as independent and identically distributed Chi-squared distribution, and the PDF of γ_1 can be expressed as

$$f_{\gamma_1}(\gamma) = \frac{\gamma^{M_p N_p - 1} \exp\left(-\gamma/(\overline{\gamma}_1/M_p)\right)}{\left(\frac{\overline{\gamma}_1}{M_p}\right)^{M_p N_p} (M_p N_p - 1)!}. \tag{8}$$

In addition, the PDF of γ_2 can be defined as

$$f_{\gamma_2}(\gamma) = \frac{\gamma^{K M_s N_p - 1} \exp\left(-\gamma/(\alpha\overline{\gamma}_2/M_s)\right)}{\left(\frac{\alpha\overline{\gamma}_2}{M_s}\right)^{K M_s N_p} (K M_s N_p - 1)!}, \tag{9}$$

http://dx.doi.org/10.5772/intechopen.80073

where $\overline{\gamma}_2 = P_2/\sigma_N^2$ is the average SNR of the secondary system. Finally, the PDF of γ_τ can be written as

$$f_{\gamma_\tau}(\gamma) = \int_0^\infty (x+1) f_{\gamma_1}((x+1)\gamma) f_{\gamma_2}(x) dx. \tag{10}$$

By substituting (8) and (9) into (10), then with the help of [32, Eq. 3.351.3] and after few manipulations, PDF expression of $f_{\gamma_\tau}(\gamma)$ is given as

$$f_{\gamma_\tau}(\gamma) = \Delta \sum_{m=0}^{M_p N_p} \binom{M_p N_p}{m} (KM_s N_p + m - 1)! \left(\frac{\gamma M_p}{\overline{\gamma}_1} + \frac{M_s}{\alpha \overline{\gamma}_2}\right)^{-KM_s N_p + m}. \tag{11}$$

Furthermore, collecting constant terms in (11), Δ is defined by

$$\Delta = \beta \gamma^{M_p N_p - 1} \exp\left(-\frac{M_p \gamma}{\overline{\gamma}_1}\right). \tag{12}$$

Hereby, β is constituted as

$$\beta = \frac{\left(\frac{\overline{\gamma}_1}{M_p}\right)^{-M_p N_p} \left(\frac{\alpha \overline{\gamma}_2}{M_s}\right)^{-KM_s N_p}}{(M_p N_p - 1)!(KM_s N_p - 1)!}. \tag{13}$$

To achieve the closed-form expression of (11), binomial expression of $\left(\frac{\gamma M_p}{\overline{\gamma}_1} + \frac{M_s}{\alpha \overline{\gamma}_2}\right)^{-KM_s N_p + m}$ term must be completed. The binomial expansion of this negative exponential term is given as

$$\left(\frac{\gamma M_p}{\overline{\gamma}_1} + \frac{M_s}{\alpha \overline{\gamma}_2}\right)^{-\zeta} = \sum_{t=0}^{\infty} (-1)^t \binom{\zeta + t - 1}{t} \left(\frac{\gamma M_p}{\overline{\gamma}_1}\right)^t \left(\frac{Ms}{\alpha \overline{\gamma}_2}\right)^{\zeta + t}, \tag{14}$$

where ζ is given as $\zeta = KM_s N_p + m$. Besides, the validation of (14) is restricted via $|\frac{\gamma M_p}{\overline{\gamma}_1}| < \frac{M_s}{\alpha \overline{\gamma}_2}$ condition. Under these conditions, the closed-form expression of f_{γ_τ} is given below:

$$f_{\gamma_\tau}(\gamma) = \Delta \sum_{m=0}^{M_p N_p} \sum_{t=0}^{\infty} (-1)^t \binom{M_p N_p}{m} (\zeta - 1)! \binom{\zeta + t - 1}{t} \left(\frac{\gamma M_p}{\overline{\gamma}_1}\right)^t \left(\frac{Ms}{\alpha \overline{\gamma}_2}\right)^{\zeta + t}. \tag{15}$$

Outage probability function of the proposed MIMO system with respect to f_{γ_τ} can be expressed as

$$P_{out} = \int_0^{2^{R_{th}} - 1} f_{\gamma_\tau}(\gamma) d\gamma. \tag{16}$$

The closed-form expression for (16) can be validated with the numerical integral operation [33].

2.3. Performance evaluation

Herein, the system performance of the MIMO CR network is studied in the presence of interference leakage for Rayleigh fading channel by comparing the analytical results with computer simulations. We assumed $P_1 = P_2 = \rho$ while $\sigma_N^2 = 1$ in the performance evaluation.

In **Figure 3**, the P_{out} performance for different R_{th} values is presented. We take $\alpha = -20$ dB, $M_p = 2$, $N_p = 2$, $K = 5$, and $M_s = 1$. It can be seen from **Figure** 3 that when R_{th} is increased from 1 to 4 bits/channel, the P_{out} performance is degraded.

In **Figure 4**, the impact of the leakage coefficient, α, on the outage probability performance is depicted for $M_p = 2$, $N_p = 2$, $K = 1$, $M_s = 1$, and $R_{th} = 3$ bits/channel. As can be seen from the figure, when α is changed from -10 dB to -30 dB, the performance of the primary system is enhanced.

In **Figure 5**, α, M_p, N_p, M_s, and R_{th} are taken as -20 dB, 2, 2, 1, and 1 bits/channel, respectively. It can be observed from the figure that increasing the number of SUs decreases the outage probability performance of the primary system considerably.

In **Figure 6**, the impact of antenna diversity on the P_{out} performance is investigated for $\alpha = -10$ dB, $K = 2$, and $R_{th} = 1$ dB. It is observed from the figure that, when the number of antennas at the primary transmitter and receiver increases, the system performance enhances. Besides, the receiver diversity effect on the system performance is greater than the transmitter diversity, as expected.

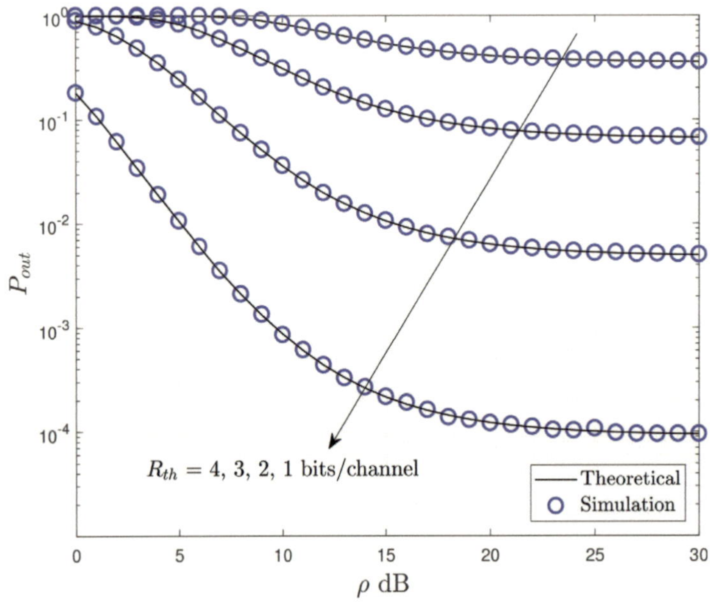

Figure 3. P_{out} performance for different data rate threshold R_{th}.

http://dx.doi.org/10.5772/intechopen.80073

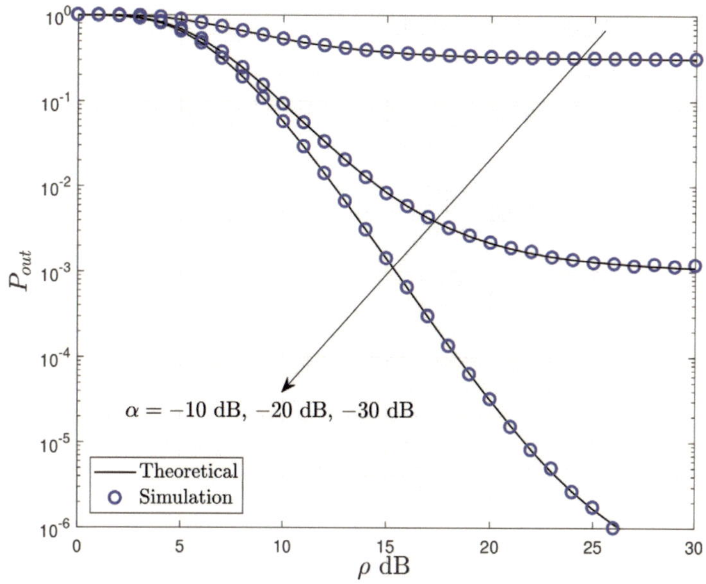

Figure 4. P_{out} performance with varying SNR for different interference-leakage values.

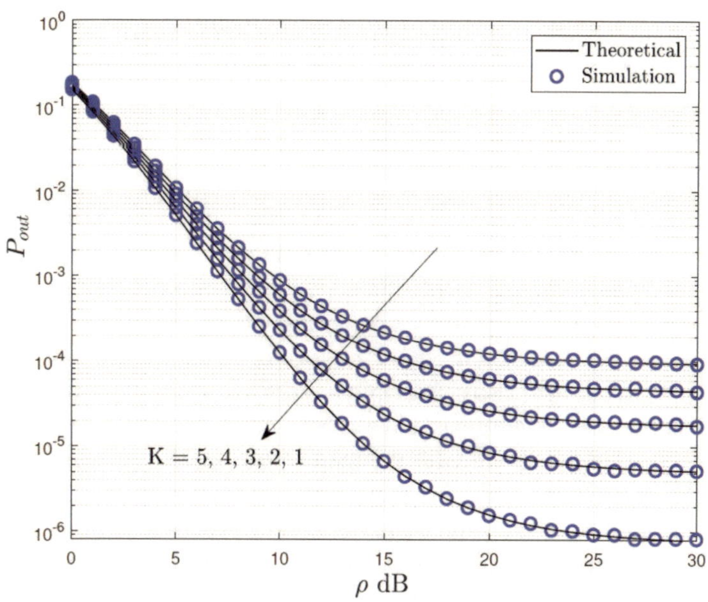

Figure 5. P_{out} vs. SNR for different numbers of SUs.

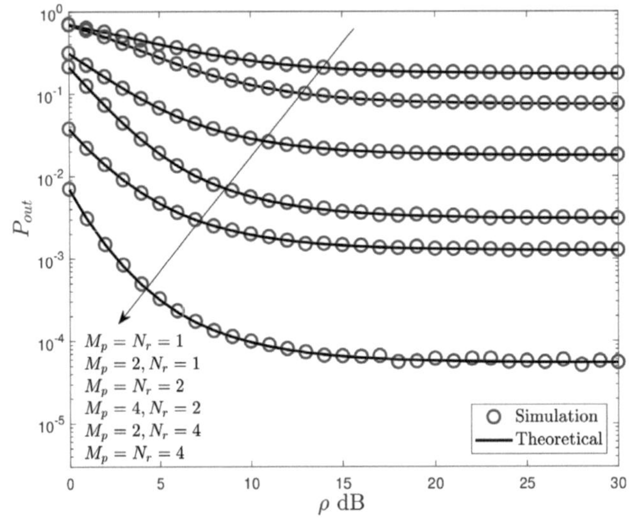

Figure 6. The effect of antenna diversity on the outage probability performance.

3. The effect of CSI quantization on interference alignment in CR networks

In this section, we investigate a cognitive two-way relaying network composed of a primary network (PN) with one pair of PU and a secondary network (SN) with two source terminals and a relay terminal (R).

3.1. System model

We consider a MIMO interference network shown in **Figure 7**, where the transmitter, T_x, and receiver, R_x, are equipped with M_1 and N_1 antennas in PN, respectively. Each PN transmitter transmits to its corresponding receiver by interfering with the SN nodes, namely, two source terminals (S_1 and S_2) and a relay terminal. That means T_x transmitter sends messages to its intended receiver R_x, whereas it also causes interference to the unintended receivers in the SN. The SN consists of two source terminals and a relay terminal. We assume that all nodes in SN operate in an AF half-duplex mode with the help of information relaying from each source terminal to R in two phases. All nodes in SN are assumed to have MIMO antennas, and there is no direct transmission between S_1 and S_2 [34–36]. We consider a scenario where the source terminals and a relay terminal are equipped with N_{S_1}, N_{S_2}, and N_R antennas, respectively. In the system model based on IA for cognitive two-way relay network, the received signal at R_x in PN can be written as

$$\mathbf{y}_{R_x} = \sqrt{\frac{P_{T_x}}{d_{R_x,T_x}^{\tau_{R_x,T_x}}}} \mathbf{U}_{R_x}^H \mathbf{H}_{R_x,T_x} \mathbf{V}_{T_x} \mathbf{s}_{T_x} + \Upsilon + \tilde{\mathbf{n}}_{R_x}, \tag{17}$$

http://dx.doi.org/10.5772/intechopen.80073

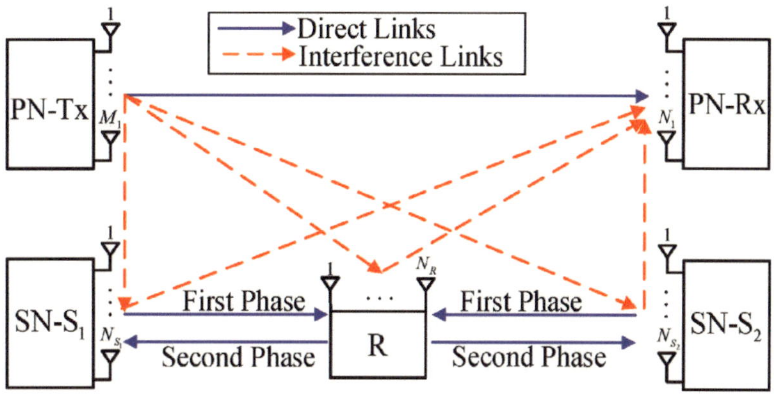

Figure 7. System model for interference alignment-based cognitive two-way relay network with primary network and secondary network.

where Υ is the interference term generated from SN to R_x defined as follows:

$$\Upsilon = \begin{cases} \sqrt{\frac{P_{S_1}}{d_{R_x,S_1}^{\tau_{R_x,S_1}}}}\mathbf{U}_{R_x}^H\mathbf{H}_{R_x,S_1}\mathbf{V}_{S_1}\mathbf{s}_{S_1} + \sqrt{\frac{P_{S_2}}{d_{R_x,S_2}^{\tau_{R_x,S_2}}}}\mathbf{U}_{R_x}^H\mathbf{H}_{R_x,S_2}\mathbf{V}_{S_2}\mathbf{s}_{S_2}, & \text{first phase} \\ \sqrt{\frac{P_R}{d_{R_x,R}^{\tau_{R_x,R}}}}\mathbf{U}_{R_x}^H\mathbf{H}_{R_x,R}\mathbf{V}_R\mathbf{s}_R, & \text{second phase.} \end{cases} \tag{18}$$

The effective additive white Gaussian noise (AWGN) term with zero mean and unit variance, $\tilde{\mathbf{n}}_{R_x}$ at R_x in PN, is defined by $\mathbf{U}_{R_x}^H\mathbf{n}_{R_x}$, where $\mathbf{n}_{R_x}$ is the AWGN vector with $\mathrm{E}\left[\mathbf{n}_{R_x}\mathbf{n}_{R_x}^H\right] = \sigma_{R_x}^2\mathbf{I}$ in which $\mathbf{I}$ is the unitary matrix, $\sigma_{R_x}^2$ is the noise variance, and $\mathrm{E}[.]$ is the expectation operator. The transmit powers at the terminals T_x, S_1, S_2, and R are denoted by P_i, for $i = T_x$, S_1, S_2, and R, respectively. Each receive node employs the interference-suppression matrix, $\mathbf{U}_j$, (for $j = R_x, R, S_1, S_2$), while each transmit node employs a precoding matrix $\mathbf{V}_i$ [37]. The conjugate transpose of the matrix is associated with the Hermitian operator $(.)^H$ [38]. The transmit signal vector for the ith user is defined by $\mathbf{s}_i$. The channel between the ith transmitter and the jth receiver nodes is denoted by $\mathbf{H}_{j,i}$ for both PN and SN. The quantized CSI is passed to the transmitter by the corresponding receiver. Because of limited feedback, the transmitters have imperfect CSI causing certain performance loss. To clarify the effect of CSI quantization error on the performance of interference alignment in underlay cognitive two-way relay networks, we investigate the BER performance, instantaneous capacity, and average PEP of the considered system. Based upon the accuracy parameter, the relation between perfect CSI ($\rho_{j,i} = 0$) and imperfect CSI ($0 < \rho_{j,i} \leq 1$) can be given as

$$\mathbf{H}_{j,i} = \sqrt{1 - \rho_{j,i}}\hat{\mathbf{H}}_{j,i} + \sqrt{\rho_{j,i}}\mathbf{E}_{j,i}, \tag{19}$$

where $\mathbf{H}_{j,i}$ is the real channel matrix and $\hat{\mathbf{H}}_{j,i}$ is the estimated channel matrix. The quantization error, $\mathbf{E}_{j,i}1mm$, can be expressed with the upper bound of $2^{-B_{j,i}/(M_1N_1-1)}$, where $B_{j,i}$ is the CSI exchange amount and M_1 and N_1 are the numbers of transmit-and-receive antennas, successively [21, 39]. It is assumed that both $\hat{\mathbf{H}}_{j,i}$ and $\mathbf{E}_{j,i}$ are independent of $\mathbf{H}_{j,i}$. Besides, each channel link is also modeled by two additional parameters: the distance between ith transmitter and the jth receiver nodes $d_{j,i}$ and the path loss exponent for the corresponding link, $\tau_{j,i}$, regarding for different radio environments, respectively.

In the first phase of the transmission (multiple-access phase) in SN, both S_1 and S_2 transmit their signals simultaneously to the relay terminal, R. Then the received signal at R can be written as

$$\mathbf{y}_R = \sqrt{\frac{P_{S_1}}{d_{R,S_1}^{\tau_{R,S_1}}}}\mathbf{U}_R^H\mathbf{H}_{R,S_1}\mathbf{V}_{S_1}\mathbf{s}_{S_1} + \sqrt{\frac{P_{S_2}}{d_{R_x,S_2}^{\tau_{R_x,S_2}}}}\mathbf{U}_R^H\mathbf{H}_{R,S_2}\mathbf{V}_{S_2}\mathbf{s}_{S_2} + \sqrt{\frac{P_{T_x}}{d_{R,T_x}^{\tau_{R,T_x}}}}\mathbf{U}_R^H\mathbf{H}_{R,T_x}\mathbf{V}_{T_x}\mathbf{s}_{T_x} + \tilde{\mathbf{n}}_R, \quad (20)$$

where $\tilde{\mathbf{n}}_R = \mathbf{U}_R^H\mathbf{n}_R$ at the relay terminal in SN is expressed as zero-mean AWGN vector with $\mathrm{E}\left[\mathbf{n}_R\mathbf{n}_R^H\right] = \sigma_R^2\mathbf{I}$ in which the noise variance at the relay terminal is depicted with σ_R^2. Besides, the received signal at S_1 and S_2 terminals in SN is defined, respectively, as

$$\mathbf{y}_{S_1} = \sqrt{\frac{P_R}{d_{S_1,R}^{\tau_{S_1,R}}}}\mathbf{U}_{S_1}^H\mathbf{H}_{S_1,R}\mathbf{V}_R\mathbf{s}_R + \sqrt{\frac{P_{T_x}}{d_{S_1,T_x}^{\tau_{S_1,T_x}}}}\mathbf{U}_{S_1}^H\mathbf{H}_{S_1,T_x}\mathbf{V}_{T_x}\mathbf{s}_{T_x} + \tilde{\mathbf{n}}_{S_1}, \quad (21)$$

$$\mathbf{y}_{S_2} = \sqrt{\frac{P_R}{d_{S_2,R}^{\tau_{S_2,R}}}}\mathbf{U}_{S_2}^H\mathbf{H}_{S_2,R}\mathbf{V}_R\mathbf{s}_R + \sqrt{\frac{P_{T_x}}{d_{S_2,T_x}^{\tau_{S_2,T_x}}}}\mathbf{U}_{S_2}^H\mathbf{H}_{S_2,T_x}\mathbf{V}_{T_x}\mathbf{s}_{T_x} + \tilde{\mathbf{n}}_{S_2}. \quad (22)$$

Here, $\tilde{\mathbf{n}}_{S_1}$ and $\tilde{\mathbf{n}}_{S_2}$ are the AWGN vector with $\mathrm{E}\left[\mathbf{n}_{S_k}\mathbf{n}_{S_k}^H\right] = \sigma_{S_k}^2\mathbf{I}$, for $k = 1, 2$ and the noise variance of $\sigma_{S_k}^2$. In addition to that, in the second phase of the signal transmission (broadcast phase), R broadcasts the combined signal $\mathbf{y}_R$ after multiplying with an ideal amplifying gain, G, which is expressed as

$$G = 1/\sqrt{\frac{P_{S_1}(1-\rho_{R,S_1})}{d_{R,S_1}^{\tau_{R,S_1}}}\|\mathbf{U}_R^H\hat{\mathbf{H}}_{R,S_1}\mathbf{V}_{S_1}\|^2 + \frac{P_{S_2}(1-\rho_{R,S_2})}{d_{R,S_2}^{\tau_{R,S_2}}}\|\mathbf{U}_R^H\hat{\mathbf{H}}_{R,S_2}\mathbf{V}_{S_2}\|^2 \ldots + \frac{P_{T_x}(1-\rho_{R,T_x})}{d_{R,T_x}^{\tau_{R,T_x}}}\|\mathbf{U}_R^H\hat{\mathbf{H}}_{R,T_x}\mathbf{V}_{T_x}\|^2}, \quad (23)$$

where $\mathbf{s}_\mathbf{R} = G\mathbf{y}_R$. We assume that both S_1 and S_2 have knowledge about their own information and can remove back-propagating self-interference from the imposed signals. We also assume that all interference at the receive terminals are perfectly aligned and the following feasible conditions are satisfied for the receive nodes:

$$\mathbf{U}_j^{\mathrm{H}}\mathbf{H}_{j,i}\mathbf{V}_i\mathbf{s}_i = 0, \quad (24)$$

$$\mathrm{rank}\left(\mathbf{U}_j^{\mathrm{H}}\mathbf{H}_{j,i}\mathbf{V}_i\mathbf{s}_i\right) = f_i, \quad (25)$$

where f_i is the degree of freedom and rank (.) denotes the rank operation of a matrix. By assuming that the interference is perfectly aligned by the proposed IA algorithm, and the

http://dx.doi.org/10.5772/intechopen.80073

channel matrices are constant during the transmission, we ensure that there is no interference from the unintended transmitters and guarantee that received signal achieves f_i degrees of freedom [39]. The corresponding signal-to-interference-plus-noise ratio (SINR) for the links $T_x \to R_x$, $S_1 \to R$, and $R \to S_2$ can be derived by

$$\gamma_{T_x \to R_x} = \frac{\frac{P_{T_x}\left(1-\rho_{R_x,T_x}\right)}{d_{R_x,T_x}^{\tau_{R_x,T_x}}} \|\mathbf{U}_{R_x}^H \hat{\mathbf{H}}_{R_x,T_x} \mathbf{V}_{T_x}\|^2}{\Psi + \sigma_{\tilde{n}_{R_x}}{}^2}, \tag{26}$$

$$\gamma_{S_1 \to R} = \frac{\frac{P_{S_1}\left(1-\rho_{R,S_1}\right)}{d_{R,S_1}^{\tau_{R,S_1}}} \|\mathbf{U}_R^H \hat{\mathbf{H}}_{R,S_1} \mathbf{V}_{S_1} \mathbf{s}_{S_1}\|^2}{\frac{P_{T_x}\rho_{R,T_x}}{d_{R,T_x}^{\tau_{R,T_x}}} \|\mathbf{U}_R^H \mathbf{E}_{R,T_x} \mathbf{V}_{T_x}\|^2 + \sigma_{\tilde{n}_R}^2}, \tag{27}$$

$$\gamma_{R \to S_2} = \frac{\frac{P_R\left(1-\rho_{S_2,R}\right)}{d_{S_2,R}^{\tau_{S_2,R}}} \|\mathbf{U}_{S_2}^H \hat{\mathbf{H}}_{S_2,R} \mathbf{V}_R \mathbf{s}_R\|^2}{\frac{P_{T_x}\rho_{S_2,T_x}}{d_{S_2,T_x}^{\tau_{S_2,T_x}}} \|\mathbf{U}_{S_2}^H \mathbf{E}_{S_2,T_x} \mathbf{V}_{T_x}\|^2 + \sigma_{\tilde{n}_{S_2}}^2}, \tag{28}$$

$$\Psi = \begin{cases} \frac{P_{S_1}\rho_{R_x,S_1}}{d_{R_x,S_1}^{\tau_{R_x,S_1}}} \|\mathbf{U}_{R_x}^H \mathbf{E}_{R_x,S_1} \mathbf{V}_{S_1}\|^2 + \frac{P_{S_2}\rho_{R_x,S_2}}{d_{R_x,S_2}^{\tau_{R_x,S_2}}} \|\mathbf{U}_{R_x}^H \mathbf{E}_{R_x,S_2} \mathbf{V}_{S_2}\|^2, & \text{first phase} \\ \frac{P_R\rho_{R_x,R}}{d_{R_x,R}^{\tau_{R_x,R}}} \|\mathbf{U}_{R_x}^H \mathbf{E}_{R_x,R} \mathbf{V}_R\|^2, & \text{second phase} \end{cases} \tag{29}$$

where $\mathbf{E}_{j,i}$ is the quantization error and $\|.\|$ is the Euclidean norm. In here, $\gamma_{S_2 \to R}$ and $\gamma_{R \to S_1}$ can be found by changing the subscript S_1 with S_2 of (27) and S_2 with S_1 of (28). Assuming the channels are reciprocal over SN direct links, thus the channel gains for $S_1 \to R$ and $R \to S_1$ and $S_2 \to R$ and $R \to S_2$ links are identical, respectively.

3.2. Performance analysis

This section starts by the instantaneous capacity analysis of the proposed system with interference alignment in underlay cognitive two-way relay networks with CSI quantization. We then study the BER and average PEP performance.

The capacity is expressed as the expected value of the mutual information between the transmitting terminal and receiving one. In light of this fact, we consider the method developed in [29]; the instantaneous capacity in PN can be expressed as

$$C_{R_x} = \log_2\left(1 + \gamma_{T_x \to R_x}\right), \tag{30}$$

where $\gamma_{T_x \to R_x}$ is the instantaneous SINR for the corresponding link of $T_x \to R_x$. On the other hand, end-to-end capacity for the SN, based on the least strong link over two-hop transmission, is denoted as follows:

$$C_R = \frac{1}{2}\log_2\left(1+\min\left(\gamma_{S_1\to R},\gamma_{R\to S_2}\right)\right) + \frac{1}{2}\log_2\left(1+\min\left(\gamma_{S_2\to R},\gamma_{R\to S_1}\right)\right). \tag{31}$$

$\gamma_{S_1\to R}$ and $\gamma_{R\to S_2}$ are the instantaneous SINR for the $S_1 \to R$ and $R \to S_2$ links, respectively.

Average BER for binary phase shift keying (BPSK) modulation can be expressed as

$$\mathrm{BER}_j = Q\left(\sqrt{\gamma_j}\right) \tag{32}$$

where $Q(x)$ is the Gaussian Q-function and defined by $Q(x) = \left(1/\sqrt{2\pi}\right)\int_x^{\infty} e^{-t^2/2}dt$ [37].

Average pairwise error probability ($\overline{\mathrm{PEP}}$) can be computed as averaging the Gaussian Q-function over Rayleigh fading statistics [40], $f_{\gamma_{T_x\to R_x}}(\gamma) = \left(e^{-\gamma/\overline{\gamma}_{T_x\to R_x}}\right)/\overline{\gamma}_{T_x\to R_x}1mm$, where $\overline{\gamma}_{T_x\to R_x} = P_{T_x}\left(1-\rho_{R_x,T_x}\right)/d_{R_x,T_x}^{\tau_{R_x,T_x}}\sigma_{\tilde{n}_{R_x}}^2$

$$\overline{\mathrm{PEP}} = \int_0^{\infty} Q\left(\sqrt{\gamma_{T_x\to R_x}}\right) f_{\gamma_{T_x\to R_x}}(\gamma)d\gamma. \tag{33}$$

Finally, this integral can be evaluated with the help of Mathematica and average PEP under Rayleigh fading channel can be derived in a closed form as follows:

$$\overline{\mathrm{PEP}} = \frac{1}{2}\left(1-\sqrt{\frac{\overline{\gamma}_{T_x\to R_x}}{2+\overline{\gamma}_{T_x\to R_x}}}\right). \tag{34}$$

3.3. Numerical results

In this section the numerical results are provided with various scenarios to evaluate the performance analysis for IA in underlay cognitive two-way relay networks with CSI quantization. BER performance for direct transmission links of the proposed system is illustrated in **Figure 8** over Rayleigh distribution for different amounts of CSI exchange with varying SNR. For convenience, we set $d_{j,i} = 3\ m$ and $\tau = 2.7$, and 3×3 MIMO configuration is studied in this figure. Because of the number of interfering links, the quantization error for the $T_x \to R_x$ transmission is greater than the other links ($S_1 \rightleftarrows R \rightleftarrows S_2$). Even if the analyzed BER performance of the SN seems better than the PN, it should not be forgotten that SN operates in half-duplex mode. Performance loss in BER due to imperfect CSI ($B_{j,i} = 4$, for instance) becomes larger as SNR increases compared to the perfect CSI (for $B_{j,i} = \infty$) case.

In **Figure 9**, the average PEP versus SNR is plotted for $d_{j,i} = 3\ m$ and $\tau = 2.7$ over Rayleigh fading channel in PN. It can be noticed from the figure that as SNR increases, average PEP decreases, as expected. To reach the perfect CSI case, we take $B_{j,i} = \infty$, and the average PEP performance noticeably enhances. We also consider the case of imperfect CSI ($B_{j,i} = 4$) for the comparison purposes in the same figure.

http://dx.doi.org/10.5772/intechopen.80073

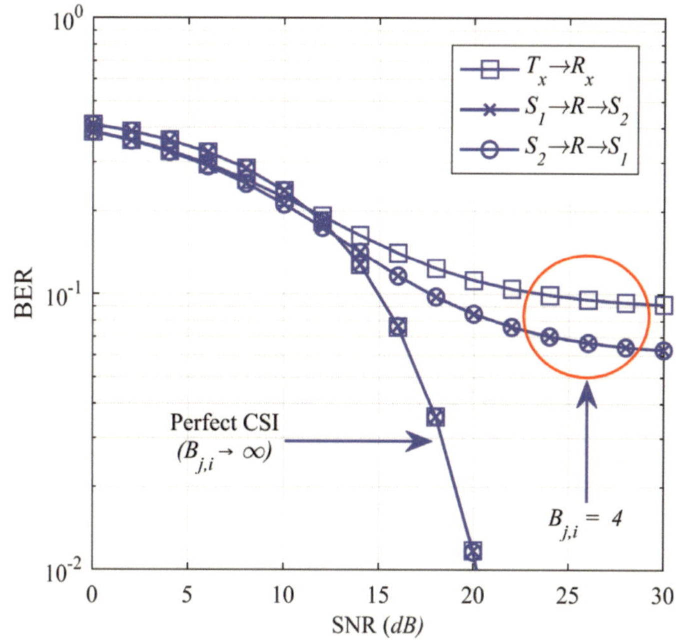

Figure 8. BER performance for different amounts of CSI exchange with varying SNR.

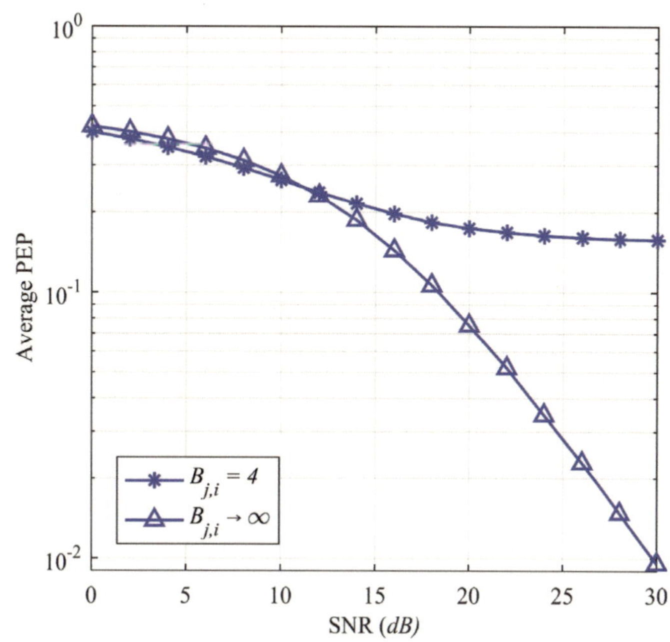

Figure 9. Average PEP performance for different amounts of CSI exchange with varying SNR over Rayleigh fading channel in primary network.

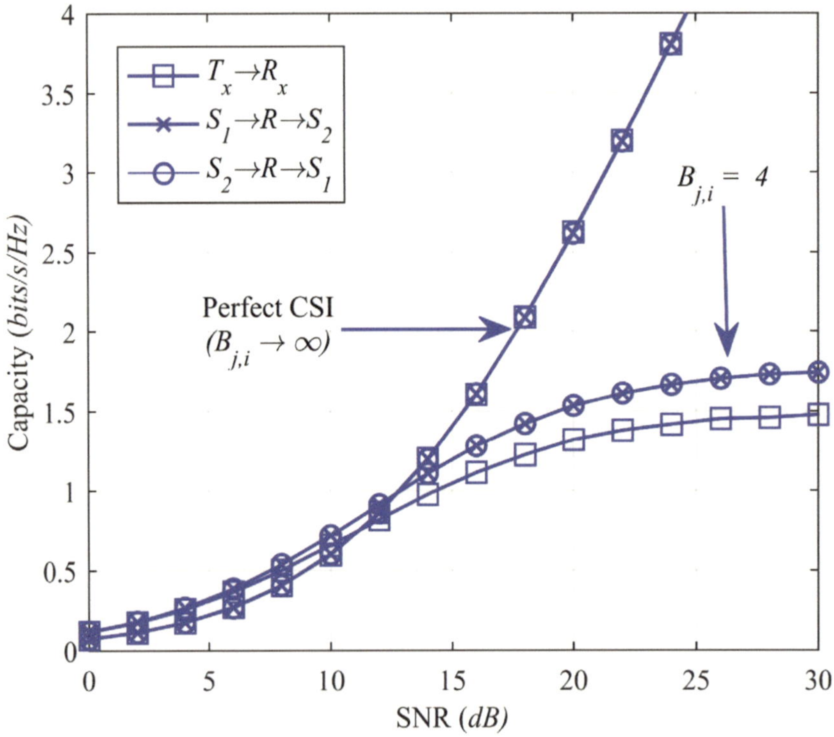

Figure 10. Capacity vs. SNR of the primary network and secondary network nodes under different CSI scenarios.

Figure 10 examines the capacity analysis with perfect and imperfect CSI for different direct links in PN and SN. The results clearly show that, examining the capacity with perfect CSI, performance improvement becomes larger as the SNR increases.

Figure 11 demonstrates the effects of $B_{j,i}$ and $d_{j,i}$ parameters on the BER performance for the SN with varying SNR when $\tau = 2.7$ and 3×3 MIMO scheme is used. The results clearly show that for a fixed SNR value, the performance of the considered system increases with the decrease of the $d_{j,i}$. It can be seen from the same figure that the increase on the amount of CSI exchange $B_{j,i}$ positively affects the BER performance.

Figure 12 shows the capacity performance of PU in the underlay cognitive two-way relay network over Rayleigh fading channel with varying path loss exponent, τ. The results show a performance improvement while the value of τ decreases. In this plot, $B_{j,i} = 8$, $d_{j,i} = 3\ m$, and the 3×3 MIMO scheme are considered. Depending on the environmental conditions for mobile communications, typical τ values, ranging from 1.6 to 5, are used to plot this figure. First, for the line of sight in a building, the environment is considered with the τ values of 1.6 and 1.8. Second, capacity is computed for the free-space environment with $\tau = 2$. Then, the capacity performance is presented with τ values of 2.7 and 3.3 for urban area cellular radio environment. Finally, the shadowed urban cellular radio environment is associated with two different τ values of 3 and 5 to analyze the capacity performance with varying SNR [41].

http://dx.doi.org/10.5772/intechopen.80073

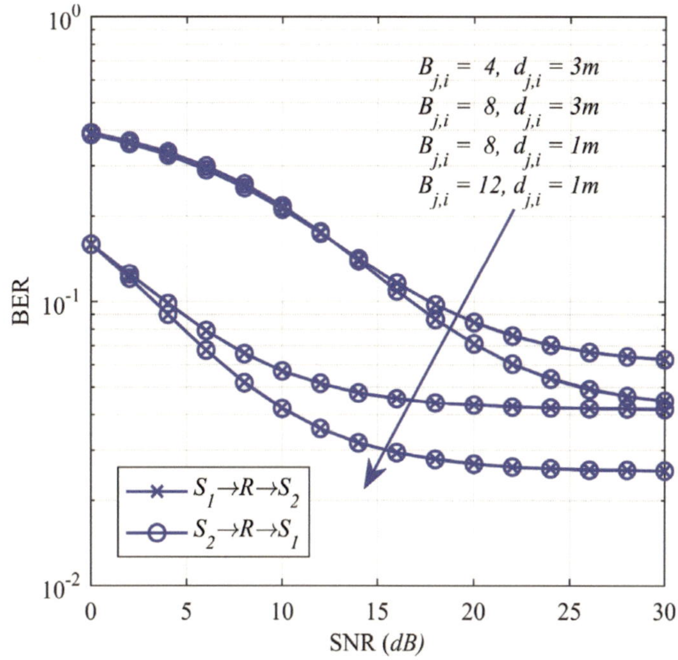

Figure 11. BER performance for different amounts of CSI exchange and distances with varying SNR over Rayleigh fading channel for secondary network.

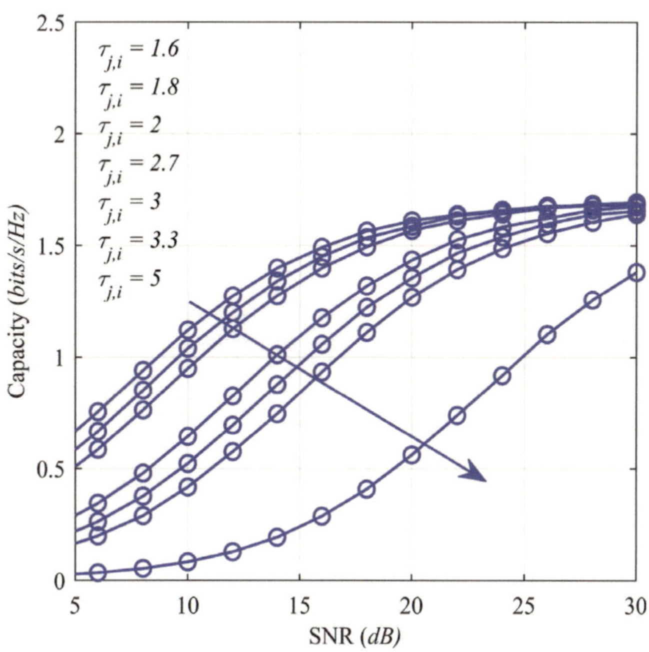

Figure 12. Capacity changes with SNR for the environmental conditions having different path loss exponents.

4. Conclusion

In this chapter, the system performance of linear interference alignment on the MIMO CR network is investigated under interference leakage. To quantify the performance of the primary system under a certain level of interference leakage, the closed-form outage probability expression is derived for Rayleigh fading channel. In all analyses, the theoretical results closely match with the simulations which confirm the accuracy of the derived expressions.

In the second part of this work, considering a practical issue, we investigate the performance of interference alignment in underlay cognitive radio network with CSI quantization error over general MIMO interference channel. Amplify-and-forward scheme for two-way relay network under Rayleigh fading is considered. The impact of the CSI exchange amount, the distance between the ith transmitter and the jth receiver nodes, and the path loss exponent on the BER performance, system capacity, and average PEP for the proposed system model are analyzed. We provide the exact closed-form expression for the average PEP in primary network over Rayleigh distribution, while IA algorithm perfectly eliminates the interference. The present performance analysis can be extended to the multiple secondary user pairs, and this approach will be another subject of our future work.

It would be interesting to study on various scenarios, including single-hop, multi-hop, and multi-way networks in future work to analyze the system performance over the recently developed interference alignment algorithms for next-generation 5G wireless communication systems.

Acknowledgements

The authors wish to express their special thanks to Seda Ustunbas (Wireless Communication Research Laboratory, Istanbul Technical University, Turkey) for useful discussions of this chapter.

Author details

Arif Basgumus[1], Mustafa Namdar[1], Hakan Alakoca[2], Eylem Erdogan[3] and Lutfiye Durak-Ata[2]*

*Address all correspondence to: lutfiye@ieee.org

1 Department of Electrical and Electronics Engineering, Dumlupinar University, Turkey

2 Informatics Institute, Applied Informatics Department, Istanbul Technical University, Turkey

3 Department of Electrical and Electronics Engineering, Istanbul Medeniyet University, Turkey

References

[1] Panwar N, Sharma S, Singh AK. A survey on 5G: The next generation of mobile communication. Physical Communication. Mar. 2016;**18**(2):64-84

[2] Wang CX et al. Cellular architecture and key technologies for 5G wireless communication networks. IEEE Communications Magazine. Feb. 2014;**52**(2):122-130

[3] Agiwal M, Roy A, Saxena N. Next generation 5G wireless networks: A comprehensive survey. IEEE Communication Surveys and Tutorials. Feb. 2016;**18**(3):1617-1655

[4] Catak E, Durak-Ata L. Waveform design considerations for 5G wireless networks. Towards 5G Wireless Networks-A Physical Layer Perspective, InTech. Dec. 2016;**18**(3):27-48

[5] Sharma SK et al. Cognitive radio techniques under practical imperfections: A survey. IEEE Communication Surveys and Tutorials. Jul. 2015;**17**(4):1858-1884

[6] Hong X, Wang J, Wang CX, Shi J. Cognitive radio in 5G: A perspective on energy-spectral efficiency trade-off. IEEE Communications Magazine. Jul. 2014;**52**(7):46-53

[7] Zhang Z, Zhang W, Zeadally S, Wang Y, Liu Y. Cognitive radio spectrum sensing framework based on multi-agent architecture for 5G networks. IEEE Wireless Communications. Dec. 2015;**22**(6):34-39

[8] Kakalou I, Psannis KE, Krawiec P, Badea R. Cognitive radio network and network service chaining toward 5G: Challenges and requirements. IEEE Communications Magazine. Nov. 2017;**55**(11):145-151

[9] Namdar M, Basgumus A. Outage performance analysis of underlay cognitive radio networks with decode-and-forward relaying. In: Cognitive Radio. InTech; 2017. pp. 25-37

[10] Razaviyayn M, Sanjabi M, Luo Z-Q. Linear transceiver design for interference alignment: Complexity and computation. IEEE Transactions on Information Theory. May 2012;**58**(5): 2896-2910

[11] Liu T, Yang C. On the feasibility of linear interference alignment for MIMO interference broadcast channels with constant coefficients. IEEE Transactions on Signal Processing. May 2013;**61**(9):2178-2191

[12] Gonzalez O, Beltran C, Santamaria I. A feasibility test for linear interference alignment in MIMO channels with constant coefficients. IEEE Transactions on Information Theory. Mar. 2014;**60**(3):1840-1856

[13] Razaviyayn M, Gennady L, Zhi-Quan L. On the degrees of freedom achievable through interference alignment in a MIMO interference channel. IEEE Transactions on Signal Processing. Feb. 2012;**60**(2):812-821

[14] Zhao N, Yu FR, Sun H, Li M. Adaptive power allocation schemes for spectrum sharing in interference-alignment-based cognitive radio networks. IEEE Transactions on Vehicular Technology. May 2016;**65**(5):3700-3714

[15] Sam PP, Gowindaswarny UM. Antenna selection and power allocation for IA-based underlay CR. IET Signal Processing. Jun. 2017;**11**(6):734-742

[16] Sultana R, Sarkar MZI, Hossain MS. Linear precoding techniques in enhancing security of cognitive radio networks. IEEE ICECTE, Rajshahi, Bangladesh. Dec. 2016:1-4

[17] Zhao N, Yu FR, Sun H. Adaptive energy-efficient power allocation in green interference-alignment-based wireless networks. IEEE Transactions on Vehicular Technology. 2015; **64**(9):4268-4281

[18] Amir M, Keyi AE, Nafie M. Constrained interference alignment and the spatial degrees of freedom of MIMO cognitive networks. IEEE Transactions on Information Theory. May 2011;**57**(5):2994-3004

[19] Jafar SA. Interference alignment: A new look at signal dimensions in a communication network. Foundations and Trends in Communications and Information Theory. Jun. 2011; **7**(1):1-134

[20] Zhao N et al. Interference alignment and its applications: A survey, research issues, and challenges. IEEE Communication Surveys and Tutorials. Mar. 2016;**18**(3):1779-1803

[21] Chen X, Yuen C. On interference alignment with imperfect CSI: Characterizations of outage probability, ergodic rate and SER. IEEE Transactions on Vehicular Technology. Jan. 2016;**65**(1):47-58

[22] Thukral J, Bolcskei H. Interference alignment with limited feedback. IEEE ISIT, Seoul, Korea. Jul. 2009:1759-1763

[23] Song B, Hardt M. Effects of imperfect channel state information on achievable rates of precoded multi-user MIMO broadcast channels with limited feedback. IEEE ICC, Deresden, Germany. Jun. 2009:1-5

[24] Rezaee M, Guillaud M. Interference alignment with quantized grassmannian feedback in the K-user constant MIMO interference channel. IEEE Transactions on Wireless Communications. Feb. 2016;**15**(2):1456-1468

[25] Krishnamachari RT, Varanasi MK. Interference alignment under limited feedback for MIMO interference channels. IEEE Transactions on Signal Processing. Aug. 2013;**61**(15): 3908-3914

[26] Sam RP, Govindaswamy UM. Antenna selection and adaptive power allocation for IA-based underlay CR. IET Signal Processing. Jun. 2017;**11**(6):734-742

[27] Tang J, Lambotharan S, Pomeroy S. Interference cancellation and alignment techniques for multiple-input and multiple-output cognitive relay networks. IET Signal Processing. Feb. 2013;**7**(3):188-200

[28] Arzykulov S, Nauryzbayev G, Tsiftsis TA, Abdallah M. Error performance of wireless powered cognitive relay networks with interference alignment. IEEE PIMRC, Montreal, Canada. Jun. 2017:1-5

[29] Arzykulov S, Nauryzbayev G, Tsiftsis TA, Abdallah M. On the capacity of wireless powered cognitive relay network with interference alignment. IEEE PIMRC, Singapore. Dec. 2017:1-6

[30] Ata SO, Altunbas I. Analog network coding over cascaded fast fading Rayleigh channels in the presence of self-interference. IEEE SIU, Zonguldak, Turkey. May 2016:253-256

[31] Chiani M, Win MZ, Zanella A. On the capacity of spatially correlated MIMO Rayleigh-fading channels. IEEE Transactions on Information Theory. Oct. 2003;**49**(10):2363-2371

[32] Gradshteyn IS, Ryzhik IM. Table of Integrals, Series, and Products. 7th ed. Amsterdam: Elsevier/Academic Press; 2007

[33] Ghavami H, Moghaddam SS. Outage probability of device to device communications underlaying cellular network in Suzuki fading channel. IEEE Communications Letters. May 2017;**21**(5):1203-1206

[34] Toan HV, Bao VNQ, Le KN. Performance analysis of cognitive underlay two-way relay networks with interference and imperfect channel state information. EURASIP Journal on Wireless Communications and Networking. Nov. 2018;**2018**(53):1-10

[35] Zhang C, Ge J, Li J, Rui Y, Guizani M. A unified approach for calculating the outage performance of two-way AF relaying over fading channels. IEEE Transactions on Vehicular Technology. Mar. 2015;**64**(3):1218-1229

[36] Nasir A, Zhou X, Durrani S, Kennedy R. Relaying protocols for wireless energy harvesting and information processing. IEEE Transactions on Wireless Communications. Jul. 2013; **12**(7):3622-3636

[37] Sung H, Park SH, Lee KJ, Lee I. Linear precoder designs for *K*-user interference channels. IEEE Transactions on Wireless Communications. Jan. 2010;**9**(1):291-301

[38] Jindal N. MIMO broadcast channels with finite-rate feedback. IEEE Transactions on Information Theory. Nov. 2006;**52**(11):5045-5060

[39] Zhao N, Yu FR, Sun H, Li M. Adaptive power allocation schemes for spectrum sharing in interference-alignment-based cognitive radio networks. IEEE Transactions on Vehicular Technology. May 2016;**65**(5):3700-3714

[40] Afana A, Mahady IA, Ikki S. Quadrature spatial modulation in MIMO cognitive radio systems with imperfect channel estimation and limited feedback. IEEE Transactions on Communications. Mar. 2017;**65**(3):981-991

[41] Rappaport TS. Wireless Communications: Principles and Practice. 2nd ed. New Jersey: Prentice Hall; 2001